A-Z
MATHEMATICS

A-Z MATHEMATICS

Praveen Prakash

CENTRUM PRESS
NEW DELHI-110002 (INDIA)

CENTRUM PRESS
H.O.: 4360/4, Ansari Road, Daryaganj,
New Delhi-110 002 (India)
Ph.: 23278000, 23261597

B.O.: No. 1015, Ist Main Road, BSK IIIrd Stage
IIIrd Phase, IIIrd Block,
Bangalore - 560 085 (India)
Tel.: 080-41723429
Visit us at: www.centrumpress.com

A-Z Mathematics

First Edition, 2009

ISBN 978-93-80106-42-7

PRINTED IN INDIA

Printed at Salasar Imaging Systems, Delhi-110035 (India)

Contents

Preface *vii*

1. Number System 1
2. Percentage and Measurement 46
3. Exponents and Radicals 67
4. Logarithms 100
5. Fundamentals of Algebra 137
6. Factoring Polynomials 166
7. Linear Equations 187
8. Complex Numbers 230
9. Quadratic Equations 249
10. Numerical Trigonometry 276

Index 298

Contents

Preface

Mathematics is the science of measurement, of establishing quantitative relationships between the properties of entities. The entities being measured occupy the whole spectrum of abstractness, from first-level concepts, which are based on perceptual data obtained by direct observation, to high-level concepts, which are further up in the edifice of knowledge. Furthermore, being the science of measurement, mathematics provides the logical glue that cements and cross-connects the structural components of the edifice.

Mathematics as a subject is the one invented by humans, for our own convenience to represent things and to do calculations, unlike other science whose basic laws are found in nature. It truly is beautiful that the abstract concepts of mathematics can be used to model the world around us. This is astounding because the laws of mathematics were not created with the universe, but have been defined by mankind over the centuries. These laws model the world so well, that people often fail to distinguish between the real situation and the mathematical model that is being used to study it. Since the world is so complex, mathematical models cannot accurately model every detail of the universe. However, they may come amazingly close and help illuminate many of the mysteries of the universe.

Author

Chapter 1

Number System

NUMBER SET

The word "set" implies a collection or grouping of similar objects or symbols. The objects in a set have at lea& one characteristic in common, such as similarity of appearance or purpose. A set of tools would be an example of a group of objects not necessarily similar in appearance but similar in purpose. The objects or symbols in a set are called members or elements of the set.

The elements of a mathematical ret are usually symbols, ouch as numerals, lines, or points. For example, the integer6 greater than zero and less than 6 form a set, as follows:

{1, 2, 3, 4}

Notice that braces are used to indicate sets. This is often done where the elements of the set are not too numerous, Since the elements of the set (2, 4, 6) are the same as the elements of (4, 2, 6}, there two seta are said to be equal. In other words, equality between sets has nothing to do with the order in which the elements are arranged. Further- more, repeated elements are not necessary. That is, the elements of (2, 2, 3, 4) are simply 2, 3, and 4. Therefore the sets (2, 3, 4) and (2, 2, 3,4) are equal.
Practice problems:

1. Use the correct symbols to designate the set of odd positive integers greater than 0 and less than 10.

2. of name of days of the week which do not contain the letter "s".
3. List the elements of the set of natural numbers greater than 15 and less than 20.
4. Suppose that we have sets as follows:

$$A = (1, 2, 3)\ C = (1, 2, 3, 4)$$
$$B = (1, 2, 2, 3)\ D = (1, 1, 2, 3)$$

Which of these sets are equal?

1. 1, 3, 5, 7, 9
2. {Monday, Friday
3. 16, 17, 18, and 19
4. A = B = D

SUBSETS

Since it is inconvenient to enumerate all of the elements of a set each time the set is mentioned, sets are often designated by a letter. For example, we could let *S* represent the set of all integers greater than 0 and less than 10. In symbols, this relationship could be stated as follows:

$$s = (1, 2, 3, 4, 5, 6, 7, 8,9)$$

Now suppose that we have another set, *T*, which comprises all positive even integers less than 10. This set is then defined as follows:

$$T - (2, 4, 6, 8)$$

Notice that every element of *T* is also an element of *S*. This establishes the SUBSET relationship; *T* is said to be a subset of 9.

POSITIVE INTEGERS

The mart fundamental ret of numbers is the set of positive integers. This ret comprises the counting numbers (natural numbers) and includes, as SUbsets, all of the sets of numbers which we have di8cueeed. The set of natural numbers has an outstanding characteristic: it is infinite. This means that the successive elements of the set continue to increase in size without limit, each number being larger by 1 than the number preceding it. Therefore there is no 'largest" number; any number that we might choose as larger than all others could be increased to a larger number simply by adding 1 to it.

One way to represent the set of natural numbers symbolically would be as follows:

(1, 2, 3, 4, 5, 6,...)

The three dots, called ellipsis, indicate that the pattern established by the numbers shown continues without limit. In other words, the next number in the set is understood to be 7, the next after that is 8, etc.

POINTS AND LINES

In addition to the many sets which can be formed with number symbols, we frequently find it necessary in mathematics to work with sets composed of points or lines. A point is an idea, rather than a tangible object, just as a number is. The mark which is made on a piece of paper is merely a symbol representing the Point. In strict mathematical terms, a point has no dimensions (physical size) at all. Thus a Pencil dot is only a rough picture of a point, useful for indicating the location of the point but certainly not to be confused with the idea.

Now suppose that a large number of points are placed side by side to form a "string." Picturing this arrangement by drawing dots on paper, we would have a "dotted line." If more dots were placed between the dots already in the string, with the number of dots increasing until we could not see between them, we would have a rough picture of a line. Once again, it is important to emphasize that the picture is only a symbol which represents an ideal line. The ideal line would have length but no width or thickness.

The foregoing discussion leads to the conclusion that a line is actually a set of points. The number of elements in the set is infinite, since the line extends in both directions without limit. The idea of arranging points together to form a line may be extended to the formation of planes (flat surfaces). A mathematical plane is determined by three points which do not lie on the same line. It is also determined by two intersecting lines.

Line Segments and Rays

When we draw a "line," label its end Points *A* and *B*, and call it 'line *AB*," we really mean LINE SEGMENT *AB*. A line segment is a subset of the set of points comprising a line. men a line is considered to have a starting point but no stopping point (that is, it extends without limit in one direction), it is called a RAY. A ray is not a line segment, because it does not terminate at both ends; it may be appropriate to refer to a ray as a 'half-line."

THE NUMBER LINE

As in the case of a line segment, a ray is a subset of the set of Points comprising a line. All three-lines, line segments, and rays-are subsets of the set of points comprising a plane, Among the many devices used for representing a set of numbers, one of the most useful is the number line. To illustrate the construction of a number line, let us place the elements of the set of natural numbers in one-to-one correspondence with points on a line. Since the natural numbers are equally spaced, we select points such that the distances between them are equal. The starting point is labeled 0, the next point is labeled 1, the next 2, etc., using the natural numbers in normal counting order. Such an arrangement is often referred to as a scale, a familiar example being the scale on a thermometer.

Thus far in our discussion, we have not mentioned any numbers other than integers. The number line is an ideal device for picturing the

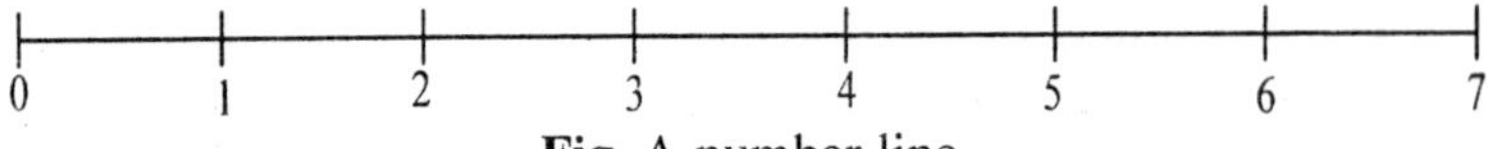

Fig. A number line

relationship between integers and other numbers such as

fractions and decimals. It is clear that many points, other than those representing integers, exist on the number line. Examples are the points representing the numbers l/2 (located halfway between 0 and 1) and 2.5 (located halfway between 2 and 3).

An interesting question arises, concerning the "in-between" points on the number line:

How many points (numbers) exist between any two integers? To answer this question, suppose that we first locate the point halfway between 0 and 1, which corresponds to the number l/2. Then let us locate the point halfway between 0 and l/2, which corresponds to the number l/4. The result of the next such halving operation could be l/8, the next l/16, etc. If we need more space to continue our halving operations on the number line, we can enlarge our "picture" and then continue.

It soon becomes apparent that the halving process could continue indefinitely; that is, without limit. In other words, the number of points between 0 and 1 is infinite. The same is true of any other interval on the number line.

Thus, between any two integers there is an infinite set of numbers other than integers. If this seems physically impossible, considering that even the sharpest pencil point has some width, remember that we are working with ideal points, which have no physical dimensions whatsoever. Although it is beyond the scope of this course to discuss such topics as orders of infinity, it is interesting to note that the set of integers contains many subsets which are themselves infinite.

Not only are the many subsets of numbers other than integers infinite, but also such subsets as the set of all odd integers and the set of all even integers. By intuition we see that these two subsets are infinite, as follows: If we select a particular odd or even integer which we think is the largest possible, a larger one can be formed immediately by merely adding 2.

Perhaps the most practical use for the number line is in explaining the meaning of negative numbers. Negative numbers are discussed in detail in chapter 3 of this course.

POSITIVE INTEGERS

The purpose of this chapter is to review the methods of combining integers. We have already used one combination process in our discussion of counting. We will extend the idea of counting, which is nothing more than simple addition, to develop a systematic method for adding numbers of any size, We will also learn the meaning of subtraction, multiplication, and division.

ADDITION AND SUBTRACTION

In the following discussion, it is assumed that the reader knows the basic addition and subtraction tables, which present such facts as the following: 2 + 3 = 5,9 + 8 = 17, 8 – 3 = 5, etc. The operation of addition is indicated by a plus sign (+) as in 8 + 4 = 12.

The numbers 8 and 4 are Addends and the answer (12) is their SUM. The operation of subtraction is indicated by a minus sign (–) as in 9 – 3 = 6. The number 9 is the Minuend, 3 is the Subtrahend, and the answer (6) is their Difference.

REGROUPING

Addition may be performed with the addends arranged horizontally, if they are small enough and not too numerous. However, the most common method of arranging the addends is to place them in vertical columns.

In this arrangement, the units digits of all the addends are a lined vertically, as are the tens digits, the hundreds digits, etc. The following example shows three addends arranged properly for addition:

$$\begin{array}{r} 357 \\ 1{,}845 \\ 22 \\ \hline \end{array}$$

It is customary to draw a line below the last addend, placing the answer below this line. Subtraction problems are arranged in columns in the same manner as for addition, with a line at the bottom and the answer below this line.

Carry and Borrow

Problems involving several addends, with two or more digits each, usually produce sums in one or more of the columns which are greater than 9. For example, suppose that we perform the following addition:

$$\begin{array}{r} 357 \\ 845 \\ 22 \\ \hline 1{,}224 \end{array}$$

The answer was found by a process called "carrying." In this process extra digits, generated when a column sum exceeds 9, are carried to the next column to the left and treated as addends in that column. Carrying may be explained by grouping the original addends. For example, 357 actually means 3 hundreds plus 5 tens plus 7 units. Rewriting the problem with each addend grouped in terms of units, tens, etc., we would have the following:

$$\begin{array}{rrr} 300+ & 50+ & 7 \\ 800+ & 40+ & 5 \\ & 20+ & 2 \\ \hline 1{,}100+ & 110+ & 14 \end{array}$$

The "extra" digit in the units column of the answer represents 1 ten. We regroup the columns of the answer so that the units column has no digits representing tens, the tens column has no digits representing hundreds, etc., as follows:

$$\begin{aligned} 1{,}100 + 110 + 14 &= 1{,}100 + 110 + 10 + 4 \\ &= 1{,}100 + 120 + 4 \\ &= 1{,}200 + 20 + 4 \\ &= 1{,}000 + 200 + 20 + 4 = 1{,}224 \end{aligned}$$

When we carry the 10 from the expression 10 + 4 to the tens column and place it with the 110 to make 120, the result is the same as if we had added 1 to the digits 5, 4, and 2 in the tens column of the original problem.

Therefore, the thought process in addition is as follows: Add the 7, 5, and 2 in the units column, getting a sum of 14.

Write down the 4 in the units column of the answer and carry the 1 to the tens column.

Mentally add the 1 along with the other digits in the tens column, getting a sum of 12. Write down the 2 in the tens column of the answer and carry the 1 to the hundreds column. Mentally add the 1 along with the other digits in the hundreds column, getting a sum of 12.

Write down the 2 in the hundreds column of the answer and carry the 1 to the thousands column. If there were other digits in the thou- sands column to which the 1 could be added, the process would continue as before. Since there are no digits in the thousands column of the original problem, this final 1 is not added to anything, but is simply written in the thousands place in the answer.

The borrow process is the reverse of carrying and is used in subtraction. Borrowing is not necessary in such problems as 46 – 5 and 58 – 53. In the first problem, the thought process may be "5 from 6 is 1 and bring down the 4 to get the difference, 41." In the second problem, the thought process is "3 from 8 is 5" and "5 from 5 is zero," and the answer is 5. More explicitly, the subtraction process in these examples is as follows:

$$\begin{array}{r} 40+6 \\ 5 \\ \hline 40+1 \end{array} = 41 \qquad \begin{array}{r} 50+8 \\ 50+3 \\ \hline 0+5 \end{array} = 5$$

This illustrates that we are subtracting units from units and tens from tens. Now consider the following problem where borrowing is involved:

$$\begin{array}{r} 43 \\ 8 \\ \hline \end{array}$$

If the student uses the borrowing method, he may think "8 from 13 is 5 and bring down 3 to get the difference, 35." In this case what actually was done is as follows:

$$\begin{array}{r} 30+13 \\ 8 \\ \hline 30+5 \end{array} = 35$$

A 10 has been borrowed from the tens column and combined with the 3 in the units column to make a number large enough for subtraction of the 8.

Notice that borrowing to increase the value of the digit in the units column reduces the value of the digit in the tens column by 1. Sometimes it is necessary to borrow in more than one column. For example, suppose that we wish to subtract 2,345 from 5,234. Grouping the minuend and subtrahend in units, tens, hundreds, etc., we have the following:

$$\begin{array}{l} 5{,}000 + 200 + 30 + 4 \\ \underline{2{,}000 + 300 + 40 + 5} \end{array}$$

Borrowing a 10 from the 30 in the tens column, we regroup as follows:

$$\begin{array}{l} 5{,}000 + 200 + 30 + 14 \\ \underline{2{,}000 + 300 + 40 + .5} \end{array}$$

In the final regrouping, we borrow from the thousands column to make subtraction possible in the hundreds column, with the following result:

$$\begin{array}{l} 5{,}000 + 200 + 30 + 14 \\ \underline{2{,}000 + 300 + 40 + \ 5} \\ 2{,}000 + 800 + 80 + 9 = 2{,}889 \end{array}$$

In actual practice, the borrowing and regrouping are done mentally. The numbers are written in the normal manner, as follows:

$$\begin{array}{r} 5{,}234 \\ \underline{-2{,}345} \\ 2{,}889 \end{array}$$

The following thought process is used: Borrow from the tens column, making the 4 become 14. Subtracting in the units column, 5 from 14 is 9. In the tens column, we now have a 2 in the minuend as a result of the firs-t borrowing operation. Some students find it helpful at first to cancel any digits that are reduced as a result of borrowing, jotting down the digit of next lower value just above the canceled digit. This has been done in the following example:

$$\begin{array}{r} 4,12 \\ -5,234 \\ -2,345 \\ \hline 2,889 \end{array}$$

After canceling the 3, we proceed with the subtraction, one column at a time. We borrow from the hundreds column to change the 2 that we now have in the tens column into 12. Subtracting in the tens column, 4 from 12 is 8. Proceeding in the same way for the hundreds column, 3 from 11 is 8. Finally, in the thousands column, 2 from 4 is 2.

Mental Calculation

Mental regrouping can be used to avoid the necessity of writing down some of the steps, or of rewriting in columns, when groups of one-digit or two-digit numbers are to be added or subtracted.

One of the most common devices for rapid addition is recognition of groups of digits whose sum is 10. For example, in the following problem two "ten groups" have been marked with braces:

$$\begin{array}{l} 7 \\ \left.\begin{array}{l} 6 \\ 4 \end{array}\right\} 10 \\ 5 \\ \left.\begin{array}{l} 1 \\ 9 \end{array}\right\} 10 \\ \hline \end{array}$$

To add this column as grouped, you would say to yourself, "7, 17, 22, 32." The thought should be just the successive totals as shown above and not such cumbersome steps as "7 + 10, 17, + 5, 22, + 10, 32."

When successive digits appear in a column and their sum is less than 10, it is often convenient to think of them, too, as a sum rather than separately.

Thus, if adding a column in which the sum of two successive digits is 10 or less, group them as follows:

$$\left.\begin{matrix}3\\1\\1\end{matrix}\right\}5$$

$$\left.\begin{matrix}8\\1\end{matrix}\right\}9$$

$$\left.\begin{matrix}4\\6\end{matrix}\right\}10$$

The thought process here might be, as shown by the grouping, "5, 14, 24." Multiplication may be indicated by a multiplication sign (x) between two numbers, a dot.

SUBTRACTION

In an.example such as 73 - 46, the conventional approach is to place 46 under 73 and subtract units from units and tens from tens, and write only the difference without the intermediate steps. To do this, the best method is to begin at the left. Thus, in the example 73 - 46, we take 40 from 73 and then take 6 from the result. This is done mentally, however, and the thought would be "73, 33, 27," or "33, 27." In the example 84 - 21 the thought is "64,63" and in the example 64 - 39 the thought is "34, 25."

MULTIPLICATION AND DIVISION

Between two numbers, or parentheses around one or both of the numbers to be multiplied. The following examples illustrate these methods:

$$6 \times 8 = 48$$

$$6 - 8 = 48$$

$$6\ (8) = 48$$

$$(6)\ (8) = 48$$

Notice that when a dot is used to indicate multiplication, it is distinguished from a decimal point or a period by being placed above the line of writing, as in example 2, whereas a period or decimal point appears on the line.

Notice also that when parentheses are used to indicate multiplication, the numbers to be multiplied are spaced closer together than they are when the dot or $\times$ is used.

In each of the four examples just given, 6 is the Multipier and 8 is the Multplicand.

Both the 6 and the 8 are Factors, and the more modern texts refer to them this way. The "answer" in a multiplication problem is the Product; in the examples just given, the product is 48.

Division usually is indicated either by a division sign (+) or by placing one number over another number with a line between the numbers, as in the following examples:

1. $8 \div 4 = 2$
2. $\frac{8}{4} = 2$

The number 8 is the Dividend, 4 is the Divisor, and 2 is the Quotient.

MULTIPLICATION METHODS

The multiplication of whole numbers may be thought of as a short process of adding equal numbers. For example, 6(5) and 6 × 5 are read as six 5's. Of course we could write 5 six times and add, but if we learn that the result is 30 we can save time. Although the concept of adding equal numbers is quite adequate in explaining multiplication of whole numbers, it is only a special case of a more general definition, which will be explained later in multiplication involving fractions.

Grouping

Let us examine the process involved in multiplying 6 times 2'7 to get the product 162. We first arrange the factors in the following manner:

$$\begin{array}{r} 27 \\ \times 6 \\ \hline 162 \end{array}$$

The thought process is as follows:

1. 6 times'7 is 42. Write down the 2 and carry the 4.
2. 6 times 2 is 12. Add the 4 that was carried over from step 1 and write the result, 16, beside the 2 that was written in step 1.
3. The final answer is 162.

Table 2.1 shows that the factors were grouped in units, tens, etc. The multiplication was done in three steps: Six times 7 units is 42 units (or 4 tens and 2 units) and six times 2 tens is 12 tens (or 1 hundred and 2 tens). Then the tens were added and the product was written as 162.

Table: 7 Multiplying by a One-digit Number

	Hundreds	Tens	Units
6(27) = 162		2	7 6
	1	4 2	2
	1	6	2

In preparing numbers for multiplication as in table 2.1, it is important to place the digits of the factors in the proper columns; that is, units must be placed in the units column, tens in tens column, and hundreds in hundreds column. Notice that it is not necessary to write the zero in the case of 12 tens (120) since the 1 and 2 are written in the proper columns. In practice, the addition is done mentally, and just the product is written without the intervening steps.

Multiplying a number with more than two digits by a one-digit number, as shown in table 2-2, involves no new ideas.

Three times 6 units is 18 units (1 ten and 8 units), 3 times 0 tens is 0, and 3 times 4 hundreds is 12 hundreds (1 thousand and 2 hundreds). Notice that it is not necessary to write the O's resulting from the step "3 times 0 tens is 0." The two terminal O's of the number 1,200 are also omitted, since the 1 and the 2 are placed in their correct columns by the position of the 4.

Table: Multiplying a Three-digit Number by a One-digit Number

	Thousands	Hundreds	Tens	Units
3(406) = 1,218		4	0	6 3
		 1	1 2	8
	1	2	1	8

Partial Products

In the example, 6(8) – 48, notice that the multiplying could be done another way to get the correct product as follows:

$$6\,(3 + 5) = 6 \times 3 + 6 \times 5$$

That is, we can break 8 into 3 and 5, multiply each of these by the other factor, and add the partial products. This idea is employed in multiplying by a two-digit number. Consider the following example:

$$\begin{array}{r} 43 \\ \times 27 \\ \hline 1,161 \end{array}$$

Breaking the 27 into 20 + 7, we have 7 units times 43 plus 2 tens times 43, are follows:

$$43(20 + 7) = (43) + (43)(20)$$

Since 7 units times 43 is 301 units, and 2 tens times 43 is 86 tens, we have the following:

$$\begin{array}{rl} 43 & \\ \times\ 27 & \\ \hline 301 & = 3 \text{ hundreds}, 0 \text{ tens}, 1 \text{ unit} \\ 86 & \\ \hline 1{,}161 & = 8 \text{ hundreds}, 6 \text{ tens} \end{array}$$

As long as the partial products are written in the correct columns, we can multiply beginning from either the left or the right of the multiplier. Thus, multiplying from the left, we have

$$\begin{array}{r} 43 \\ \times 27 \\ \hline 86 \\ 301 \\ \hline 1{,}161 \end{array}$$

Multiplication by a number having more places involves no new ideas.

End Zeros

The placement of partial products must be kept in mind when multiplying in problems involving end zeros, as in the following example:

$$\begin{array}{r} 27 \\ \times 40 \\ \hline 1{,}080 \end{array}$$

We have 0 units times 27 plus 4 tens times 27, as follows:

$$\begin{array}{r} 27 \\ \times 40 \\ \hline 0 \\ 106 \\ \hline 1{,}060 \end{array}$$

The zero in the units place plays an important part in the reading of the final product. End zeros are often called "place holders" since their only function in the problem is to hold the digit positions which they occupy, thus helping to place the other digits in the problem correctly. The end zero in the foregoing problem can be accounted for very nicely, while at the same time placing the other digits correctly, by means of a shortcut. This consists of offsetting the 40 one place to the right and then simply bringing down the 0, without using it as a multiplier at all. The problem would appear as follows:

$$\begin{array}{r} 27 \\ \times 40 \\ \hline 1,080 \end{array}$$

If the problem involves a multiplier with more than one end 0, the multiplier is offset as many places to the right as there are end 0's. For example, consider the following multiplication in which the multiplier, 300, has two end O's:

$$\begin{array}{r} 220 \\ \times 300 \\ \hline 66,000 \end{array}$$

Notice that there are as many place-holding zeros at the end in the product as there are place-holding zeros in the multiplier and the multiplicand combined.

Placement of Decimal Points

In any whole number in the decimal system, there is understood to be a terminating mark, called a decimal point, at the right-hand end of the number. Although the decimal point is seldom shown except in numbers involving decimal fractions (covered in chapter 5 of this course), its location must be known. The placement of the decimal point is automatically taken care of when the end O's are correctly placed. Practice problems. Multiply in each of the following problems:

1. 187 × 8
2. 67 × 49
3. 940 × 20
4. 807 × 28
5. 694 × 80
6. 9,241 × 7,800

DIVISION METHODS

Just as multiplication can be considered as repeated addition, division can be considered as repeated subtraction. For example, if we wish to divide 12 by 4 we may subtract 4 from 12 in successive steps and tally the number of times that the subtraction is performed, as follows:

$$\begin{array}{r} 12 \\ 4^{*} \\ \hline 8 \\ 4^{*} \\ \hline 4 \\ 4^{*} \\ \hline 0 \end{array}$$

As indicated by the asterisks used as tally marks, 4 has been subtracted 3 times. This result is sometimes described by saying that "4 is contained in 12 three times."

Since successive subtraction is too cumbersome for rapid, concise calculation, methods which treat division as the inverse of multiplication are more useful. Knowledge of the multiplication tables should lead us to an answer for a problem such as 12 + 4 immediately, since we know that 3 × 4 is 12.

However, a problem such as 84 + 4 is not so easy to solve by direct reference to the multiplication table.

One way to divide 84 by 4 is to note that 84 is the same as 80 plus 4. Thus 84 + 4 is the same as 80 + 4 plus 4 + 4. In symbols, this can be indicated as follows:

$$\begin{array}{r} 20+1 \\ 4\overline{)80+4} \end{array}$$

(When this type of division symbol is used, the quotient is written above the vinculum as shown.) Thus, 84 divided by 4 is 21.

From the foregoing example, it can be seen that the regrouping is useful in division as well as in multiplication.

However, the mechanical procedure used in division does not include writing down the regrouped form of the dividend. After becoming familiar with the process, we find that the division can be performed directly, one digit at a time, with the regrouping taking place mentally. The following example illustrates this:

$$\begin{array}{r} 14 \\ 4\overline{)56} \\ 4 \\ \hline 16 \\ \underline{16} \end{array}$$

The thought process is as follows: "4 is contained in 5 once" (write 1 in tens place over the 5); "one times 4 is 4" (write 4 in tens place under 5, take the difference, and bring down 6); and "4 is contained in 16 four times" (write 4 in units place over the 6).

After a little practice, many people can do the work shown under the dividend mentally and write only the quotient, if the divisor has only 1 digit. The divisor is sometimes too large to be contained in the first digit of the dividend. The following example illustrates a problem of this kind:

$$\begin{array}{r} 36 \\ 7\overline{)252} \\ 21 \\ \hline 42 \\ 42 \\ \hline \end{array}$$

Since 2 is not large enough to contain 7, we divide 7 into the number formed by the first two digits, 25. Seven is contained 3 times in 25; we write 3 above the 5 of the dividend. Multiplying, 3 times 7 is 21; we write 21 below the first two digits of the dividend.

Subtracting, 25 minus 21 is 4; we write down the 4 and bring down the 2 in the units place of the dividend. We have now formed a new dividend, 42.

Seven is contained 6 times in 42; we write 6 above the 2 of the dividend. Multiplying as before, 6 times 7 is 42; we write this product below the dividend 42. Subtracting, we have nothing left and the division is complete.

Estimation

When there are two or more digits in the divisor, it is not always easy to determine the first digit of the quotient. An estimate must be made, and the resulting trial quotient may be too large or too small. For example, if 1,862 is to be divided by 38, we might estimate that 38 is contained 5 times in 186 and the first digit of our trial divisor would be 5. However, multiplication reveals that the product of 5 and 38 is larger than 186. Thus we would change the 5 in our quotient to 4, and the problem would then appear as follows:

```
       49
38 ) 1862
     152
     ----
      342
      342
      ---
```

On the other hand, suppose that we had estimated that 38 is contained in 186 only 3 times. We would then have the following:

```
        3
38 ) 1862
     114
     ---
      72
```

Now, before we make any further moves in the division process, it should be obvious that something is wrong. If our new dividend is large enough to contain the divisor before bringing down a digit from the original dividend, then the trial quotient should have been larger. In other words, our estimate is too small.

Proficiency in estimating trial quotients is gained through practice and familiarity with number combinations. For example, after a little experience we realize that a close estimate can be made in the foregoing problem by thinking of 38 as "almost 40." It is easy to see that 40 is contained 4 times in 186, since 4 times 40 is 160. Also, since 5 times 40 is 200, we are reasonably certain that 5 is too large for our trial divisor.

Uneven Division

In some division problems such as 7 ÷ 3, there is no other whole number that, when multiplied by the divisor, will give the dividend. We use the distributive idea to show how division is done in such a case. For example, 7 ÷ 3 could be written as follows:

Thus, we see that the quotient also carries one unit that is to be divided by 3. It should now be clear that $3\overline{)37} = 3\overline{)30+7}$. and that this can be further reduced as follows:

$$\frac{30}{3}+\frac{6}{3}+\frac{1}{3}=10+2+\frac{1}{3}=12\frac{1}{3}$$

In elementary arithmetic the part of the dividend that cannot be divided evenly by the divisor is often called a REMAINDER and is placed next to the quotient with the prefix *R*. Thus, in the foregoing example where the quotient was $12\frac{1}{3}$, the quotient could be written 12 *R* 1. This method of indicating uneven division is useful in examples such as the following:

Suppose that $13 is available for the purchase of spare parts, and the parts needed cost $3 each. Four parts can be bought with the available money, and $1 will be left over. Since it is not possible to buy 1/3 of a part, expressing the result as 4 R 1 gives a more meaningful answer than 4 1/3.

Placement of Decimal Points In division, as in multiplication, the placement of the decimal point is important. Determining the location of the decimal point and the number of places in the quotient can be relatively simple if the work is kept in the proper columns. For example, notice the vertical alignment in the following problem:

$$\begin{array}{r} 311 \\ 31\overline{)9,641} \\ 9\,3 \\ \hline 34 \\ 31 \\ \hline 31 \\ 31 \\ \hline \end{array}$$

We notice that the first two places in the dividend are used to obtain the first place in the quotient. Since 3 is in the hundreds column there are two more places in the quotient (tens place and units place).

The decimal point in the quotient is understood to be directly above the position of the decimal point in the dividend. In the example shown here, the decimal point is not shown but is understood to be immediately after the second 1.

Checking Accuracy. The accuracy of a division of numbers can be checked by multiplying the quotient by the divisor and adding the remainder, if any. The result should equal the dividend. Consider the following example:

```
       5203
  42)218541              5203
     210               ×   42
     ---               ------
      85                10406
      84                20812
      ---              ------
      141       Check: 218526
      126                + 15
      ---              ------
       15              218541
```

DENOMINATE NUMBERS

We have learned that denominate numbers are not difficult to add and subtract, provided that units, tens, hundreds, etc., are retained in their respective columns. Multiplication and division of denominate numbers may also be performed with comparative ease, by using the experience gained in addition and subtraction.

Multiplication

In multiplying denominate numbers by integers, no new ideas are needed. If in the problem 3(5 yd 2 ft 6 in.) we remember that we can multiply each part separately to get the correct product (as in the example, 6(8) = 6(3) + 6(5)), we can easily find the product, as follows:

```
5 yd 2 ft 6 in
           ×3
--------------
15 yd ft 18 in.
```

Simplifying, this is 17 yd 1 ft 6 in.

When one denominate number is multiplied by another, a question arises concerning the products of the units of measurement. The product of one unit times another of the same kind is one square unit.

For example, 1 ft times 1 ft is 1 square foot, abbreviated sq ft; 2 in. times 3 in. is 6 sq in.; etc. If it becomes necessary to multiply such numbers as 2 yd 1 ft times 6 yd 2 ft, the foot units may be converted to fractions of a yard, as follows:

$$(2 \text{ yd } 1 \text{ ft})(6 \text{ yd } 2 \text{ ft}) = (2\ 1/3 \text{ yd})(6\ 2/3 \text{ yd})$$

In order to complete the multiplication, a knowledge of fractions is needed. Fractions are discussed in chapter 4 of this training course.

Division

The division of denominate numbers requires division of the highest units first; and if there is a remainder, conversion to the next lower unit, and repeated division until all units have been divided.

In the example (24 gal 1 qt 1 pt) + 5, we perform the following steps:

Step 1:
$$\begin{array}{l} \quad\ \ 4 \text{ gal} \\ 5\overline{)24 \text{ gal}} \\ \quad\ 20 \\ \quad\ \overline{\ \ 4} \text{ gal (left over)} \end{array}$$

Step 2: Convert the 4 gal left over to 16 qt and add to the 1 qt.

Step 3:
$$\begin{array}{l} \quad\ \ 3 \text{ qt} \\ 5\overline{)17 \text{ qt}} \\ \quad\ 15 \\ \quad\ \overline{\ \ 2} \text{ qt (left over)} \end{array}$$

Step 4: Convert the 2 qt left over to 4 pt and add to the 1 pt.

Step 5: $5\overset{1\,pt}{\overline{)5\,qt}}$

Therefore, 24 gal 1 qt 1 pt. Divided by 5 is 4 gal. 3 qt 1 pt.

Practice problem. In problem8 1 through 4, divide as indicated. In problems 5 through 8, multiply or divide as indicated.

1. 549 + 9
2. 470/63
3. $25\sqrt{2,300}$
4. $64\sqrt{74,516}$
5. $\begin{array}{r} 4\,hr\,26\,min\,16\,sec \\ \times 5 \\ \hline \end{array}$
6. 3(4 gal 3 qt 1 pt)
7. $\dfrac{67\,deg\,43\,min\,12\,sec}{2}$
8. $5\sqrt{63\,lb\,11\,oz}$

ORDER OF OPERATIONS

When a series of operations involving addition, subtraction, multiplication, or division is indicated, the order in which the operations are performed is important only if division is involved or if the operations are mixed. A series of individual additions, subtractions, or multiplication may be performed in any order.

Thus, In

$$4 + 2 + 7 + 5 = 18$$

or

$$100 - 20 - 10 - 3 = 67$$

or

$$4 \times 2 \times 7 \times 5 = 280$$

the numbers may be combined in any order desired. For

example, they may be grouped easily to give

$$6 + 12 = 18$$

$$97 - 30 = 67$$

$$40 \times 7 = 280$$

A series of divisions should be taken in the order written. Thus,

$$100 \div 10 \div 2 = 10 \div 2 = 5$$

In a series of mixed operations, perform multiplications and divisions in order from left to right, then perform additions and subtractions in order from left to right.

For example

$$100 \div 4 \times 5 = 25 \times 5 = 125$$

and

$$60 - 25 \div 5 = 60 - 5 = 55$$

Now consider

$$\begin{aligned} & 60 - 25 + 5 + 15 - 100 + 4 \times 10 \\ &= 60 - 5 + 15 - 100 + 4 \times 10 \\ &= 60 - 5 + 15 - 100 + 40 \\ &= 115 - 105 = 10 \end{aligned}$$

Practice problems. Evaluate each of the following expressions:

1. $9 \div 3 + 2$
2. $18 - 2 \times 5 + 4$
3. $90 \div 2 \div 9$
4. $75 \div 5 \times 3 \div 5$
5. $7 + 1 - 8 \times 4 \div 16$

MULTIPLES AND FACTORS

Any number that is exactly divisible by a given number is a Multiple of the given number. For example, 24 is a multiple of 2, 3, 4, 6, 8, and 12, since it is divisible by each of these numbers. Saying that 24 is a multiple of 3, for instance, is equivalent to saying that 3 multiplied by some whole number will give 24. Any number is a multiple of itself and also of 1.

Any number that is a multiple of 2 is an even number. The even numbers begin with 2 and progress by 2's as follows:

2, 4, 6, 8, 10, 12,...

Any number that Is not a multiple of 2 Is an odd number. The odd numbers begin with 1 and progress by 2's, as follows:

1, 3, 5, 7, 9, 11, 13,...

Any number that CM be divided into a given number without a remainder ir a factor of the given number. The given number Is a multiple of any number that is one of Its factors. For example, 2, S, 4, 6; 8, and 12 are factors of 24. The following four equalities show various combinations of the factors of 24:

$24 = 24 \times 1 \qquad 24 = 8 \times 3$

$24 = 12 \times 2 \qquad 24 = 6 \times 4$

If the number 24 is factored as completely as possible, it assumes the form

$$24 = 2 \times 2 \times 2 \times 3$$

ZERO AS A FACTOR

If any number is multiplied by zero, the product is zero. For example, 5 times zero equals zero and may be written 5(0) = 0. The zero factor law tells us that, if the product of two or more factors is zero, at least one of the factors must be zero.

PRIME FACTORS

A number that has factors other than itself and 1 is a Composite number. For example, the number 15 is composite. It has the factors 5 and 3. A number that has no factors except itself and 1 is a prime number. Since it is sometimes advantageous to separate a composite number into prime factors, it is helpful to be able to recognize a few prime numbers quickly. The following series shows all the prime numbers up to 60:

2, 3, 5, 7, 11, 13, 17, 19, 23, 29, 31, 37, 41, 43,47,53, 59.

Notice that 2 is the only even prime number. All other even numbers are divisible by 2. Notice also that 51, for example, does not appear in the series, since it is a composite number equal to 3 × 17.

H a factor of a number Is prime, It is called a prime factor, To separate a number into prime factors, begin by taking out the smallest factor.

If the number is even, take out all the 2's first, then try 3 as a factor, etc. Thus, we have the following example:

$$\begin{aligned} 540 &= 2.\ 270 \\ &= 2.\ 2.\ 135 \\ &= 2.\ 2.\ 3.\ 45 \\ &= 2.\ 2.\ 3.\ 15 \\ &= 2.\ 2.\ 3.\ 3.\ 3.\ 5 \end{aligned}$$

Since 1 Is M understood factor of every number, we do not waste space recording It as one of the factors in a presentation of this kind, A convenient way of keeping track of the prime factor Is in the short division process as follows:

2/540

2/270

2/135

2/45

2/15

$5\frac{/5}{1}$

If a number is odd, its factors will be odd numbers. To separate an odd number into prime factors, take out the 3's first, if there are any. Then try 5 as a factor, etc. As an example,

$$\begin{aligned} 5{,}775 &= 3 \times 1,\ 925 \\ &= 3 \times 5 \times 385 \\ &= 3 \times 5 \times 5 \times 77 \\ &= 3 \times 5 \times 5 \times 7 \times 11 \end{aligned}$$

Practice Problems:

1. Which of the following are prime numbers and which are composite numbers ?

   25, 7, 18, 29, 51

2. What prime numbers are factor of 36 ?
3. Which of the following are multiplies of 3 ?.

   45, 53, 51, 39, 47

4. Find the prime factors of 27.

Answers:

1. Primes: 7, 29

   Comosite: 25, 18, 51

2. $36 = 2 \times 2 \times 3 \times 3$
3. 45, 51, 39
4. $27 = 3 \times 3 \times 3$

Tests for Divisibility

It is often useful to be able to tell by inspection whether a number is exactly divisible by one or more of the digits from 2 through 9. An expression which is frequently used, although it is sometimes misleading, is "evenly divisible." This expression has nothing to do with the concept of even and odd numbers, and it probably should be avoided in favor of the more descriptive expression, "exactly divisible." For the remainder of this discussion, the word "divisible" has the same meaning as "exactly divisible."

Several tests for divisibility are listed in the following paragraphs:

1. A number is divisible by 2 if its right-hand digit is even.
2. A number is divisible by 3 if the sum of its digits is divisible by 3. For example, the digits of the number 6,561 add to produce the sum 18. Since 18 is divisible by 3, we know that 6,561 is divisible by 3.

3. A number is divisible by 4 if the number formed by the two right-hand digits is divisible by 4. For example, the two right-hand digits of the number 3,524 form the number 24. Since 24 is divisible by 4, we know that 3,524 is divisible by 4.
4. A number is divisible by 5 if its right-hand digit is 0 or 5.
5. A number is divisible by 6 if it is even and the sum of its digits is divisible by 3. For example, the sum of the digits of 64,236 is 21, which is divisible by 3. Since 64,236 is also an even number, we know that it is divisible by 6.
6. No short method has been found for determining whether a number is divisible by 7.
7. A number is divisible by 8 if the number formed by the three right-hand digits is divisible by 8. For example, the three right-hand digits of the number 54,272 form the number 272, which is divisible by 8. Therefore, we know that 54,272 is divisible by 8.
8. A number is divisible by 9 if the sum of its digits is divisible by 9. For example, the sum of the digits of 546,372 is 27, which is divisible by 9. Therefore we know that 546,372 is divisible by 9.

Practice problems. Check each of the following numbers for divisibility by all of the digits except 7:

1. 242,431,231,320
2. 844,624,221,840
3. 988,446,662,640
4. 207,634,542,480

SIGNED NUMBERS

The positive numbers with which we have worked in previous chapters are not sufficient for every situation which may arise. For example, a negative number results in the operation of subtraction when the subtrahend is larger than the minuend.

NEGATIVE NUMBERS

When the subtrahend happens to be larger than the minuend, this fact is indicated by placing a minus sign in front of the difference, as in the following:

$$12 - 20 = -8$$

The difference, –8, is said to be negative. A number preceded by a minus sign is a negative number. The number –8 is read "minus eight." Such a number might arise when we speak of temperature changes. If the temperature was 12 degrees yesterday and dropped 20 degrees today, the reading today would be 12 – 20, or –8 degrees. Numbers that show either a plus or minus sign are called signed numbers. An unsigned number is understood to be positive and is treated as though there were a plus sign preceding it.

If it is desired to emphasize the fact that a number is positive, a plus sign is placed in front of the number, as in +5, which is read "plus five. '1 Therefore, either +5 or 5 indicates that the number 5 is positive.

If a number is negative, a minus sign must appear in.front of it, as in –9. In dealing with signed numbers it should be emphasized that the plus and minus signs have two separate and distinct functions. They may indicate whether a number is positive or negative, or they may indicate the operation of addition or subtraction.

When operating entirely with positive numbers, it is not necessary to be concerned with this distinction since plus or minus signs indicate only addition or subtraction. However, when negative numbers are also involved in a computation, it is important to distinguish between a sign of operation and the sign of a number.

DIRECTION OF MEASUREMENT

Signed numbers provide a convenient way of indicating opposite directions with a minimum of words. For example, an altitude of 20 ft above sea level could be designated as +20 ft. The same distance below sea level would then be designated as –20 ft. One of the most common devices utilizing signed numbers to indicate direction of measurement is the thermometer.

Thermometer

The Celsius (centigrade) thermometer shown in figure 3-1 illustrates the use of positive and negative numbers to indicate direction of travel above and below 0. The 0 mark is the changeover point, at which the signs of the scale numbers change from – to +. When the thermometer is heated by the surrounding air or by a hot liquid in which it is placed, the mercury expands and travels up the tube.

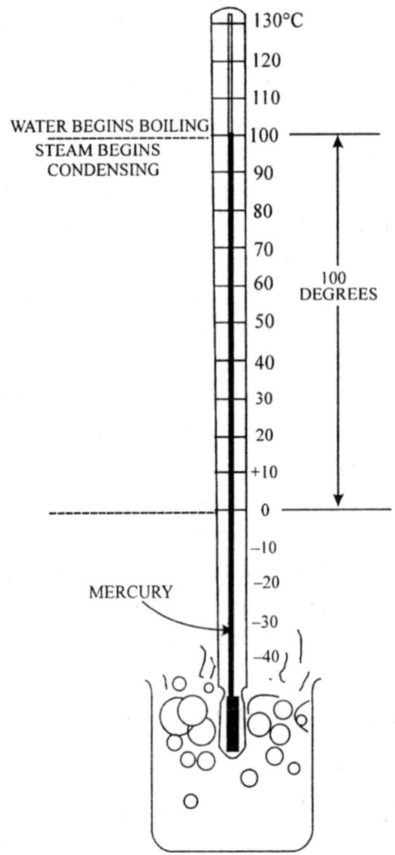

Fig. Celsius (centigrade) Temperature scale. A Measurement of 3 units below the Starting point

After the expanding mercury passes 0, the mark at which it comes to rest is read as a positive temperature. If the thermometer is allowed to cool, the mercury contracts. After passing 0 in its downward movement, any mark at which it comes to rest is read as a negative temperature.

Rectangular Coordinate System

As a matter of convenience, mathematicians have agreed to follow certain conventions as to the use of signed numbers in directional measurement. For example, in figure 3-2, a direction to the right along the horizontal line is positive, while the opposite direction (toward the left) is negative.

On the vertical line, direction upward is positive, while direction downward is negative. A distance of –3 units along the horizontal line indicates a measurement of 3 units to the left of starting point 0. A distance of -3 units on the vertical line indicates. The two lines of the rectangular coordinate system which pass through the 0 position are the vertical axis and horizontal axis. Other vertical and horizontal lines may be included, forming a grid. When such a grid is used for the location of points and lines, the resulting "picture" containing points and lines is called a GRAPH.

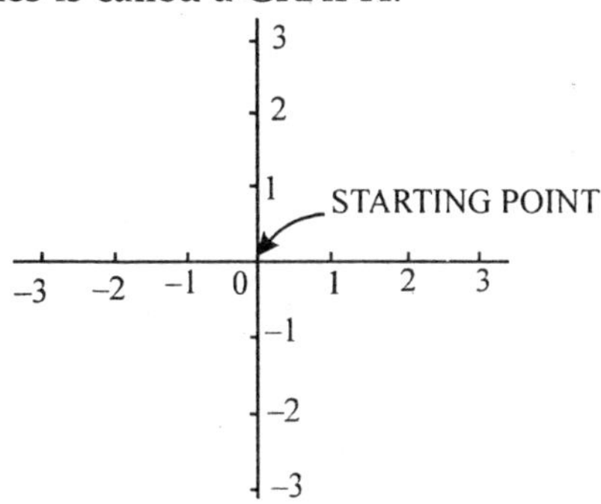

Fig. Rectangular Coordinate System

The Number Line

Sometimes it is important to know the relative greatness (magnitude) of positive and negative numbers. To determine whether a particular number is greater or less than another number, think of all the numbers both positive and negative as being arranged along a horizontal line.

–5 –4 –3 –2 –1 0 +1 +2 +3 +4 +5

Fig. Number Line Showing both Positive and Negative Numbers

Place zero at the middle of the line. Let the positive numbers extend from zero toward the right. Let the negative numbers extend from zero toward the left. With this arrangement, positive and negative numbers are so located that they progress from smaller to larger numbers as we move from left to right along the line.

Any number that lies to the right of a given number is greater than the given number. A number that lies to the left of a given number is less than the given number. This arrangement shows that any negative number is smaller than any positive number.

The symbol for "greater than" is $>$. The symbol for "less than" is $<$. It is easy to distinguish between these symbols because the symbol used always opens toward the larger number. For example, "7 is greater than 4" can be written $7 > 4$ and "–5 is less than –1" can be written $-5 < -1$.

Absolute Value

The ABSOLUTE VALUE of a number is its numerical value when the sign is dropped. The absolute value of either +5 or –5 is 5. Thus, two numbers that differ only in sign have the same absolute value. The symbol for absolute value consists of two vertical bars placed one on each side of the number, as in $|-5| = 5$. Consider also the following:

$$|4 - 20| = 16$$

$$|+7| = |-7| = 7$$

The expression $|-7|$ is read "absolute value of minus seven."

When positive and negative numbers are used to indicate direction of measurement, we are concerned only with absolute value, if we wish to know only the distance covered. For example, in figure 3-2, if an object moves to the left from the starting point to the point indicated by –2, the actual distance covered is 2 units. We are concerned only with the fact that $| -2 | = 2$, if our only interest is in the distance and not the direction.

OPERATING WITH SIGNED NUMBERS

The number line can be used to demonstrate addition of signed numbers. Two cases must be considered; namely, adding numbers with like signs and adding numbers with unlike signs.

ADDING WITH LIKE SIGNS

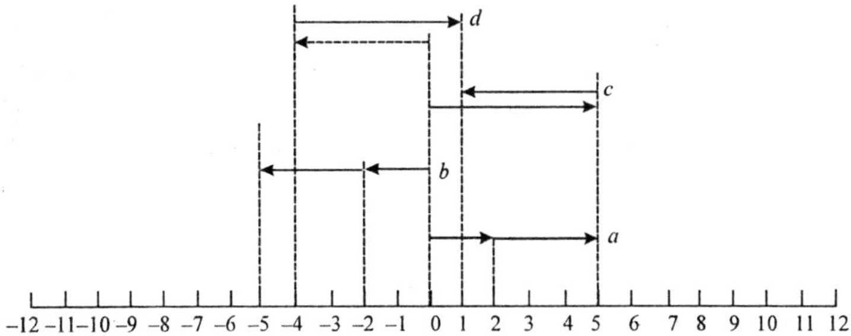

Fig. Using the number line to add. operation. Line a (fig. 3-4) above the number line shows this addition. Find 2 on the number line. To add 3 to it, go three units more in a positive direction and get 5

As an example of addition with like signs, suppose that we use the number line to add 2 + 3. Since these are signed numbers, we indicate this addition as (+2) + (+3). This emphasizes that, among the three + signs shown, two are number signs and one is a sign of To add two negative numbers on the number line, such as –2 and –3, find –2 on the number line and then go three units more in the negative direction to get –5, as in b above the number line. Observation of the results of the foregoing operations on the number line leads us to the following conclusion, which may be stated as a law: To add numbers with like signs, add the absolute values and prefix the common sign.

ADDING WITH UNLIKE SIGNS

To add a positive and a negative number, such as (–4) + (+5), find +5 on the number line and go four units in a negative direction, as in line c above the number line in figure 3-4. Notice that this addition could be performed in the other direction. That is, we could start at –4 and move 5 units in the positive direction.

The results of our operations with mixed signs on the number line lead to the following conclusion, which may be stated as a law: To add numbers with unlike signs, find the difference between their absolute values and prefix the. sign of the numerically greater number. The following examples show the addition of the numbers 3 and 5 with the four possible combinations of signs:

$$\begin{array}{cccc} 3 & -3 & 3 & -3 \\ \underline{5} & \underline{-5} & \underline{-5} & \underline{5} \\ 8 & -8 & -2 & 2 \end{array}$$

In the first example, 3 and 5 have like signs and the common sign is understood to be positive. The sum of the absolute values is 8 and no sign is prefixed to this sum, thus signifying that the sign of the 8 is understood to be positive. In the second example, the 3 and 5 again have like signs, but their common sign is negative.

The sum of the absolute values is 8, and this time the common sign is prefixed to the sum. The answer is thus –8. In the third example, the 3 and 5 have unlike signs. The difference between their absolute values is 2, and the sign of the larger addend is negative. Therefore, the answer is –2. In the fourth example, the 3 and 5 again have unlike signs. The difference of the absolute values is still 2, but this time the sign of the larger addend is positive. Therefore, the sign prefixed to the 2 is positive (understood) and the final answer is simply 2.

These four examples could be written in a different form, emphasizing the distinction between the sign of a number and an operational sign, as follows:

$$(+3) + (+5) = +8$$
$$(-3) + (-5) = -8$$
$$(+3) + (-5) = -2$$
$$(-3) + (+5) = +2$$

Practice problems: Add as indicated:

1. $-10 + 5 = (-10) + (+5) = ?$
2. Add –9, –16, and 25
3. $-7 - 1 - 3 = (-7) + (-1) + (-3) = ?$
4. Add –22 and –13

SUBTRACTION

Subtraction is the inverse of addition. When subtraction is performed, we "take away" the subtrahend. This means that whatever the value of the subtrahend, its effect is to be reversed when subtraction is indicated. In addition, the sum of 5 and –2 is 3. In subtraction, however, to take away the effect of the –2, the quantity +2 must be added. Thus the difference between +5 and –2 is +7.

Keeping this idea in mind, we may now proceed to examine the various combinations of subtraction involving signed numbers. Let us first consider the four possibilities where the minuend is numerically greater than the subtrahend, as in the following examples:

$$\begin{array}{r} 8 \\ 5 \\ \hline 3 \end{array} \quad \begin{array}{r} 8 \\ -5 \\ \hline 13 \end{array} \quad \begin{array}{r} -8 \\ 5 \\ \hline -13 \end{array} \quad \begin{array}{r} -8 \\ -5 \\ \hline -3 \end{array}$$

We may show how each of these results is obtained by use of the number line.

In the first example, we find +8 on the number line, then subtract 5 by making a movement that reverses its sign. Thus, we move to the left 5 units. The result (difference) is +3. In the second example, we find +8 on the number line, then subtract

(–5) by making a movement that will reverse its sign. Thus we move to the right 5 units. The result in this case is +13.

In the third example, we find –8 on the number line, then subtract 5 by making a movement that reverses its sign. Thus we move to the left 5 units. The result is –13.

In the fourth example, we find -8 on the number line, then reverse the sign of –5 by moving 5 units to the right. The result is –3.

Next, let us consider the four possibilities that arise when the subtrahend is numerically greater than the minuend, as in the following examples:

$$\begin{array}{r} 5 \\ 8 \\ \hline -3 \end{array} \quad \begin{array}{r} 5 \\ -8 \\ \hline 13 \end{array} \quad \begin{array}{r} -5 \\ 8 \\ \hline -13 \end{array} \quad \begin{array}{r} -5 \\ -8 \\ \hline 3 \end{array}$$

In the first example, we find +5 on the number line, then subtract 8 by making a movement that reverses its sign. Thus we move to the left 8 units. The result is –3.

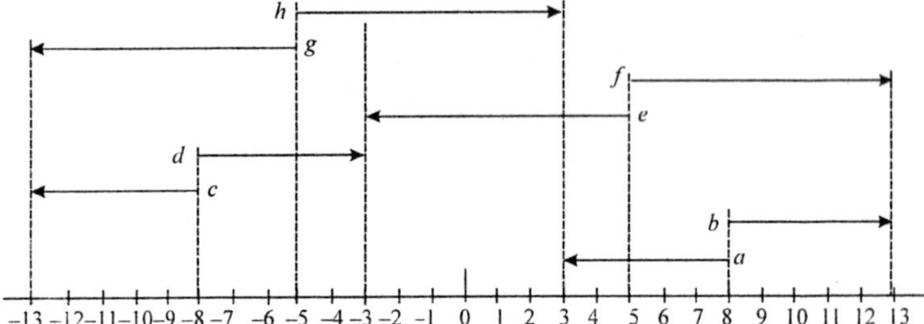

Fig. Subtraction by Use of the Number Line

In the second example, we find +5 on the number line, then subtract –8 by making a movement to the right that reverses its sign. The result is 13.

In the third example, we find –5 on the number line, then reverse the sign of 8 by a movement to the left. The result is –13.

In the fourth example, we find –5 on the number line, then reverse the sign of –8 by a movement to the right. The result is 3.

Careful study of the preceding examples leads to the

following conclusion, which is stated as a law for subtraction of signed numbers: In any subtraction problem, mentally change the sign of the subtrahend and proceed as in addition.

Practice problems. In problems 1 through 4, subtract the lower number from the upper. In 5 through 8, subtract as indicated.

1. $\begin{array}{r} 17 \\ \underline{-10} \end{array}$ 2. $\begin{array}{r} -12 \\ \underline{8} \end{array}$ 3. $\begin{array}{r} -9 \\ \underline{-13} \end{array}$ 4. $\begin{array}{r} 7 \\ \underline{16} \end{array}$

5. 1 –(–5) = ? 6. –6 –(–8) = ? 7. 14 – 7 –(–3) = ?

8. –9 – 2 = ?

MULTIPLICATION

To explain the rules for multiplication of signed numbers, we recall that multiplication of whole numbers may be thought of as shortened addition. Two types of multiplication problems must be examined; the first type involves number8 with unlike signs, and the second involves numbers with like signs.

Unlike Signs

Consider the example 3(–4), in which the multiplicand is negative. This means we are to add –4 three times; that is, 3(–4) is equal to (–4) + (–4) + (–4), which is equal to –12. For example, if we have three 4-dollar debts, we owe 12 dollars in all.

When the multiplier is negative, as in –3(7), we are to take away 7 three times. Thus, –3(7) is equal to –(7) – (7) – (7) which is equal to –21. For example, if 7 shells were expended in one firing, 7 the next, and 7 the next, there would be a loss of 21 shells in all. Thus, the rule is as follows: The product of two numbers with unlike signs is negative.

The law of signs for unlike signs is sometimes stated as follows: Minus times plus is minus; plus times minus is minus. Thus a problem such as 3(–4) can be reduced to the following two steps:

1. Multiply the signs and write down the sign of' the answer before working with the numbers themselves.

2. Multiply the numbers as if they were unsigned numbers.

Using the suggested procedure, the sign of the answer for 3(–4) is found to be minus. The product of 3 and 4 is 12, and the final answer is –12. When there are more than two numbers to be multiplied, the signs are taken in pairs until the final sign is determined.

Like Signs

When both factors are positive, as in 4(5), the sign of the product is positive. We are to add +5 four times, as follows:

$$4(5) = 5 + 5 + 5 + 5 = 20$$

When both factors are negative, as in –4(–5), the sign of the product is positive. We are to take away –5 four times.

$$-4(-5) - (5) - (-5) - (-5) = +5 + 5 + 5 = 20$$

Remember that taking away a negative 5 is the same as adding a positive 5. For example, suppose someone owes a man 20 dollars and pays him back (or diminishes the debt) 5 dollars at a time. He takes away a debt of 20 dollars by giving him four positive 5-dollar bills, or a total of 20 positive dollars in all.

The rule developed by the foregoing example is as follows: The product of two numbers with like signs is positive.

Knowing that the product of two positive numbers or two negative numbers is positive, we can conclude that the product of any even number of negative numbers is positive. Similarly, the product of any odd number of negative numbers is negative.

The laws of signs may be combined as follows: Minus times plus is minus; plus times minus is minus; minus times minus is plus; plus times plus is plus. Use of this combined rule may be illustrated as follows:

$$4(-2) - (-5) - (6) - (-3) = -720$$

Taking the signs in pairs, the understood plus on the 4 times the minus on the 2 produces a minus. This minus times the minus on the 5 produces a plus. This plus times the understood plus on the 6 produces a plus. This plus times the minus on the 3 produces a minus, so we know that the final answer is negative. The product of the numbers, disregarding their signs, is 720; therefore, the final answer is -720.

Practice problems. Multiply as indicated:

1. $5(-8) = ?$
2. $-7(3)(2) = ?$
3. $6(-1)(-4) = ?$
4. $-2(3)(-4)(5)(-6) = ?$

DIVISION

Because division is the inverse of multiplication, we can quickly develop the rules for division of signed numbers by comparison with the corresponding multiplication rules, as in the following examples:

1. Division involving two numbers with unlike signs is related to multiplication with unlike signs, as follows:

$$3(-4) = -12$$

Therefore, $\frac{-12}{3} = -4$

Thus, the rule for division with unlike signs is: The quotient of two numbers with unlike signs is negative.

2. Division involving two numbers with like signs is related to multiplication with like signs, as follows:

$$3(-4) = -12$$

Therefore, $\frac{-12}{-4} = 3$

Thus the rule for division with like signs is: The quotient of two numbers with like signs is positive.

The following examples show the application of the rules for dividing signed numbers:

1. $15 + -5$ 2. $-2(-3)) - 6$ 3. $\frac{(-3)(4)}{-6}$ 4. $-81)9$

SPECIAL CASES

Two special cases arise frequently in which the laws of signs may be used to advantage. The first such usage is in simplifying subtraction; the second is in changing the signs of the numerator and denominator when division is indicated in the form of a fraction.

Subtraction

The rules for subtraction may be simplified by use of the laws of signs, if each expression to be subtracted is considered as being multiplied by a negative sign. For example, 4 –(–5) is the same as 4 + 5, since minus times minus is plus. This result also establishes a basis for the rule governing removal of parentheses. The parentheses rule, as usually stated, is: Parentheses preceded by a minus sign may be removed, if the signs of all terms within the parentheses are changed. This is illustrated as follows:

$$12 - (3 - 2 + 4) = 12 - 3 + 2 - 4$$

The reason for the changes of sign is clear when the negative sign preceding the parentheses is considered to be a multiplier for the whole parenthetical expression. Division in Fractional Form Division is often indicated by writing the dividend as the –numerator, and the divisor as the denominator, of a fraction.

In algebra, every fraction is considered to have three signs. The numerator has a sign, the denominator has a sign, and the fraction itself, taken as a whole, has a sign. In many cases, one or more of these signs will be positive, and thus will not be shown. For example, in the following fraction the sign of the numerator and the sign of the denominator are both positive (understood) and the sign of the fraction itself is negative:

$$-\frac{4}{5}$$

Fractions with more than one negative sign are always reducible to a simpler form with at most one negative sign.

For example, the sign of the numerator and the sign of the denominator may be both negative. We note that minus divided by minus gives the same result as plus divided by plus. Therefore, we may change to the less complicated form having plus signs (understood) for both numerator and denominator, as follows:

$$\frac{-15}{-5} = \frac{+15}{+5} = \frac{15}{5}$$

Since – 15 divided by –5 is 3, and 15 divided by 5 is also 3, we conclude that the change of sign doss not alter the final answer. The same reasoning may be applied in the following ex- ample, in which the sign of the fraction itself is negative:

$$= \frac{-15}{-5} = -\frac{+15}{+5} = -\frac{15}{5}$$

When the fraction itself has a negative sign, as in this example, the fraction may be enclosed in parentheses temporarily, for the purpose of working with the numerator and denominator only. Then the sign of the fraction is applied separately to the result, as follows:

$$= \frac{-15}{-5} = -\left(\frac{-15}{-5}\right) = -(3) = -3$$

All of this can be done mentally.

If a fraction has a negative sign in one of the three sign positions, this sign may be moved to another position. Such an adjustment is an ad- vantage in some types of complicated expressions involving fractions. Examples of this type of sign change follow:

$$= \frac{15}{5} = \frac{-15}{5} = \frac{15}{-5}$$

In the first expression of the foregoing example, the sign of the numerator is positive (understood) and the sign of the fraction is negative. Changing both of these signs, we obtain the second expression. To obtain the third expression from the second we change the sign of the numerator and the sign of the denominator.

Observe that the sign changes in each case involve a pair

of signs. This leads to the law of signs for 'fractions: Any two of the three signs of a fraction may be changed without altering the value of the fraction.

AXIOMS AND LAWS

An axiom is a self-evident truth.. It is a truth that is so universally accepted that it does not require proof. For example, the statement that "a straight line is the shortest distance between two points" is an axiom from plane geometry. One tends to accept the truth of an axiom without proof, because anything which is axiomatic is, by its very nature, obviously true. On the other hand, a law (in the mathematical sense) is the result of defining certain quantities and relationships and then developing logical conclusions from the definitions.

AXIOMS OF EQUALITY

The four axioms of equality with which we are concerned in arithmetic and algebra are stated as follows:

1. If the same quantity is added to each of two equal quantities, the resulting quantities are equal. This is sometimes stated as follows: H equals are added to equals, the results are equal. For example, by adding the same quantity (3) to both sides of the following equation, we obtain two sums which are equal:

$$-2 = -3 + 1$$
$$-2 + 3 = -3 + 1 + 3$$
$$1 = 1$$

2. If the same quantity is subtracted from each of two equal quantities, the resulting quantities are equal. This is sometimes stated as follows: If equals are subtracted from equals, the results are equal. For example, by subtracting 2 from both sides of the following equation we obtain results which are equal:

$$5 = 2 + 3$$
$$5 - 2 = 2 + 3 - 2$$
$$3 = 3$$

3. If two equal quantities are multiplied by the same quantity, the resulting products are equal. This is sometimes stated as follows: If equals are multiplied by equals, the products are equal. For example, both sides of the following equation are multiplied by –3 and equal results are obtained:

$$5 = 2 + 3$$
$$(-3)\ (5) = (-3)(2 + 3)$$
$$-15 = -15$$

4. If two equal quantities are divided by the same quantity, the resulting quotients are equal. This is sometimes stated as follows:. If equals are divided by equals, the results are equal. For example, both sides of the following equation are divided by 3, and the resulting quotients are equal:

$$12 + 3 = 15$$
$$\frac{12+3}{3} = \frac{15}{3}$$
$$4 + 1 = 5$$

These axioms are especially useful when letters are used to represent numbers. If we know that $5x = -30$, for instance, then dividing both $5x$ and –30 by 5 leads to the conclusion that $x = -6$.

LAWS FOR COMBINING NUMBERS

Numbers are combined in accordance with the following basic laws:

1. The associative laws of addition and multiplication.
2. The commutative laws of addition and multiplication.
3. The distributive law.

Associative Law of Addition

The word "associative" suggests association or grouping. This law states that the sum of three or more addends is the same regardless of the manner in which they are grouped. For example, 6 + 3 + 1 is the same as 6 + (3 + 1) or (6 +3) +1. This law can be applied to subtraction by changing signs in such a way that all negative signs are treated as number signs rather than operational signs. That is, some of the addends can be negative numbers. For example, 6 – 4 – 2 can be rewritten as 6 + (–4) + (–2). By the associative law, this is the same as

$$6 + [(-4) + (-2)] \text{ or } [6 + (-4)] + (-2).$$

However, 6 – 4 – 2 is not the same as 6 – (4 – 2); the terms must be expressed as addends before applying the associative law of addition.

Associative Law of Multiplication

This law states that the product of three or more factors is the same regardless of the manner in which they are grouped. For example, 6 × 3 × 2 is the same as (6 × 3) × 2 or 6 × (3 × 2). Negative signs require no special treatment in the application of this law. For exam le 6 – (–4) – (–2) is the same as [6 –(–4)]. – (2) or 6 – [(–4) – (–2)].

Commutative Law of Addition

The word "commute" means to change, substitute or move from place to place. The commutative law of addition states that the sum of two or more addends is the same regardless of the order in which they are arranged. For example, 4 + 3 + 2 is the same as 4 + 2 + 3 or 2 + 4 + 3.

This law can be applied to subtraction by changing signs so that all negative signs become number signs and all signs of operation are positive. For example, 5 – 3 – 2 is changed to 5 + (–3) + (–2), which is the same as 5 + (–2) + (–3) or (–3) + 5 + (–2).

Commutative Law of Multiplication

This law states that the product of two or more factors is the same regardless of the order in which the factors are arranged. For example, 3 × 4 × 5 is the same as 5 × 3 × 4 or 4 ×

3 × 5. Negative signs require no special treatment in the application of this law. For example, 2 * (–4) * (–3) is the same as (–4) × (–3) × 2 or (–3) × 2 × (–4).

Distributive Law

This law.combines the operations of addition and multiplication. The word "distributive" refers to the distribution of a common multiplier among the terms of an additive expression. For example,

$$2(3 + 4 + 5) = 2.3 + 2.4 + 2.5 = 6 + 8 + 10$$

To verify the distributive law, we note that 2(3 + 4 + 5) is the same as 2(12) or 24. Also, 6 + 8 + 10 is 24. For application of the distributive law where negative signs appear, the following procedure is recommended:

$$3(4-2) = 3[4+(2-2)] = 3(4) = (4) + 3(-2) = 12 - 6 = 6$$

Chapter 2

Percentage and Measurement

In the discussion of decimal fractions, it was shown that for convenience in writing fractions whose denominators are 10 or some power of 10, the decimal point could be employed and the denominators could be dropped. Thus, this special group of fractions could be written in a much simpler way. As early as the 15th century, businessmen made use of certain decimal fractions so much that they gave them the special designation Percent.

MEANING OF PERCENT

The word "percent" is derived from Latin. It was originally "per centum," which means 'by the hundred." Thus the statement is often made that "percent means hundredths. "

Percentage deals with the group of decimal fractions whose denominators are 100-that is, fractions of two decimal places. Since hundredths were used so frequently, the decimal point was dropped and the symbol % was placed after the number and read "percent" (per 100). Thus, 0.15 and 15% represent the same value, 15/100. The first is read "15 hundredths," and the second is read "15 percent." Both mean 15 parts out of 100.

Ordinarily, percent is used in discussing relative values. For example, 25 percent may convey an idea of relative value or relationship. To say "25 percent of the crew is ashore" gives an idea of what part of the crew is gone, but it does not tell how many. For example, 25 percent of the crew would represent vastly different numbers if the comparison were made between an LSM and a cruiser. When it is necessary to use a percent in computation, the number is written in its decimal form to avoid confusion. By converting all decimal fractions so that they had the common denominator 100, men found that they could mentally visualize the relative size of the part of the whole that was being considered.

CHANGING DECIMALS TO PERCENT

Since percent means hundredths, any decimal may be changed to percent by first expressing it as a fraction with 100 as the denominator, The numerator of the fraction thus formed indicates how many hundredths we have, and therefore it indicates "how many percent" we have. For example, 0.36 is the same as 36/100. Therefore, 0.36 expressed as a percentage would be 36 percent. By the same reasoning, since 0.052 is equal to 5.2/100, 0.052 is the same as 5.2 percent.

In actual practice, the step in which the de- nominator 100 occurs is seldom written down. The expression in terms of hundredths is converted mentally to percent. This results in the following rule: To changes decimal to percent, multiply the decimal by 100 and annex the percent sign (%). Since multiplying by 100 has the effect of moving the decimal point two places to the right, the rule is sometimes stated as follows: To change a decimal to percent, move the decimal point two places to the right and annex the percent sign.

Changing Common Fractions and Whole Numbers To Percent

Common fractions are changed to percent by first expressing them as decimals. For example, the fraction 1/4 is equivalent to the decimal 0.25. Thus 1/4 is the same as 25 percent. Whole numbers may be considered as special types of decimals (for example, 4 may be written as 4.00) and thus may be expressed in terms of percentage, The meaning of an expression such as 400 percent is vague unless we keep in mind that percentage is a form of comparison. For example, a question which often arises is "How can I have more than 100 percent of something, if 100 percent means all of it?"

This question seems reasonable, if we limit our attention to such quantities as test scores. However, it is also reasonable to use percent- age in comparing a current set of data with a previous set. For example, if the amount of electrical power used by a Navy facility this year is double the amount used last year, then this year's power usage is 200 percent of last year's usage. The meaning of a phrase such as "200 percent of last year's usage" is often misinterpreted. A total amount that is 200 percent of the previous amount is not the same as an increase of 200 percent. The increase in this case is only 100 percent, for a total of 200. If the increase had been 200 percent, then the new usage figure would be 300 percent' of the previous figure.

Baseball batting averages comprise a special case in which percentage is used with only occasional reference to the word "percent." The percentages in batting averages are expressed in their decimal form, with the figure 1.000 representing 100 percent. Although a batting average of 0.300 is referred to as "batting 300," this is actually erroneous nomenclature from the strictly mathematical standpoint. The correct statement, mathematically, would be "batting point three zero zero" or "batting 30 percent."

CHANGING A PERCENT TO A DECIMAL

Since we do not compute with numbers in the percent form, it is often necessary to change a percent back to the decimal form. The procedure is just opposite to that used in changing decimals to percents: To change a percent to a decimal, drop the percent sign and divide the number by 100. Mechanically, the decimal point is simply shifted two places to the left and the percent sign is dropped. For example, 25 percent is the same as the decimal 0.25. Percents larger than 100 percent are changed to decimals by the same procedure as ordinary percents. For example, 125 percent is equivalent to 1.25.

THE THREE PERCENTAGE CASES

To explain the cases that arise in problems involving percents, it is necessary to define the terms that will be used. Rate (r) is the number of hundredths parts taken. This is the number followed by the percent sign. The base (b) is the whole on which the rate operates. Percentage (p) is the part of the base determined by the rate. In the example

$$5\% \text{ of } 40 = 2$$

5% is the rate, 40 is the base, and 2 is the percentage.

There are three cases that usually arise in dealing with percentage, as follows:

Case I. To find the percentage when the base and rate are known.

Example: What number is 6% of 50?

Case II. To find the rate when the base and percentage are known.

Example: 20 is what percent of 60?

Case III-To find the base when the percentage and rate are known.

Example: The number 5 is 25% of what number ?

Case I.

In the example

$$6\% \text{ of } 50 = ?$$

the "of" has the same meaning as it does in fractional examples, such as

$$\frac{1}{4} \text{ of } 16 = ?$$

In other words, "of" means to multiply. Thus, to find the percentage, multiply the base by the rate. Of course the rate must be changed from a percent to a decimal before multiplying can be done. Rate times base equals percentage.

Thus,

$$6\% \text{ of } 50 = 3$$

$$0.06 \times 50 = 3$$

The number that is 6% of 50 is 3.

Fractional Percents: A fractional percent represents a part of 1 percent. In a case such as this, it is sometimes easier to find 1 percent of the number and then find the fractional part. For example, we would find 1/4 percent of 840 as follows:

Therefore,

$$1\% \text{ of } 840 = 0.01 \times 840 = 8.40$$

$$\frac{1}{4}\% \text{ of } 840 = 8.40 \times \frac{1}{4} = 2.10$$

To explain case II and case II, we notice in the foregoing example that the base corresponds to the multiplicand, the rate corresponds to the multiplier, and the percentage corresponds to the product.

$$\begin{array}{r} 50(\text{base or multiplicand}) \\ .06(\text{rate or multiplier}) \\ \hline 3.00(\text{percentage or product}) \end{array}$$

Recalling that the product divided by one of its factors gives the other factor, we can solve the following problem:

$$?\ \% \text{ of } 60 = 20$$

We are given the base (60) and percentage (20).

$$\begin{array}{r} 60(\text{base}) \\ ?(\text{rate}) \\ \hline 20(\text{percentage}) \end{array}$$

We then divide the product (percentage) by the multiplicand (base) to get the other factor (rate). Percentage divided by base equals rate. The rate is found as follows:

$$\frac{20}{60} = \frac{1}{3} = .33\frac{1}{3} = 33\frac{1}{3}\%(\text{rate})$$

The rule for case II, as illustrated in the foregoing problem, is as follows: To find the rate when the percentage and base are known, divide the percentage by the base. Write the quotient in the decimal form first, and finally as a percent.

Case III.

The unknown factor in case III is the base, and the rate and percentage are known.

Example: 25% of ? = 5

$$\begin{array}{r} ?\,(\text{base}) \\ .25\ (\text{rate}) \\ \hline 5.00\,(\text{percentage}) \end{array}$$

We divide the product by its known factor to find the other factor. Percentage divided by rate equals base. Thus,

$$\frac{5}{.25} = 20(\textit{base})$$

The rule for case III may be stated as follows: To find the boss when the rate and percentage are known, divide the percentage by the rate.

PRINCIPLES OF MEASUREMENT

Computation with decimals frequently involves the addition or subtraction of numbers which do not have the same number of decimal places. For example, we may be asked to add such numbers as 4.1 and 32.31582. How should they be added? Should zeros be annexed to 4.1 until it is of the same order as the other decimal (to the same number of places)? Or, should.31582 be rounded off to tenths? Would the sum be accurate to tenths or hundred-thousandths? The answers to these questions depend on how the numbers originally arise.

Some decimals are finite or are considered as such because of their use. For instance, the decimal that represents 1/2, that is 0.5, is as accurate at 0.5 as it is at 0.5000. Likewise, the decimal that represents 1/8 has the value 0.125 and could be written just as accurately with additional end zeros. Such numbers are said to be finite. Counting numbers are finite. Dollars and cents are examples of finite values. Thus, $10.25 and $5.00 are finite values.

To add the decimals that represent 1/8 and 1/2, it is not necessary to round off 0.125 to tenths. Thus, 0.5 + 0.125 is added as follows:

$$\begin{array}{r} 0.500 \\ 0.125 \\ \hline 0.625 \end{array}$$

Notice that the end zeros were added to 0.5 to carry it out the same number of places as 0.125. It is not necessary to write such place-holding zeros if the figures are kept in the correct columns and decimal points are aligned.

Decimals that have a definite fixed value may be added or subtracted although they are of different order. On the other hand, if the numbers result from measurement of some kind, then the question of how much to round off must be decided in terms of the precision and accuracy of the measurements.

ESTIMATION

Suppose that two numbers to be added resulted from measurement. Let us say that one number was measured with a ruler marked off in tenths of an inch and was found, to the nearest tenth of an inch, to be 2.3 inches. The other number measured with a precision rule was found, to the nearest thousandth of an inch, to be 1.426 inches.

Each of these measurements requires estimation between marks on the rule, and estimation between marks on any measuring instrument is subject to human error. Experience has shown that the best the average person can do with consistency is to decide whether a measurement is more or less than halfway between marks.

The correct way to state this fact mathematically is to say that a measurement made with an instrument marked off in tenths of an inch involves a maximum probable error of 0.05 inch (five hundredths is one-half of one tenth). By the same reasoning, the probable error in a measurement made with an instrument marked in thousandths of an inch is 0.0005 inch.

PRECISION

In general, the probable error in any measurement is one-half the size of the smallest division on the measuring 'instrument. Thus the precision of a measurement depends upon how precisely the instrument is marked. It is important to realize that precision refers to the size of the smallest division on the scale; it has nothing to do with the correctness of the markings. In other words, to say that one instrument is more precise than another does not imply that the less precise instrument is poorly manufactured. In fact, it would be possible to make an instrument with very high apparent precision, and yet mark it carelessly so that measurements taken with it would be inaccurate.

From the mathematical standpoint, the precision of a number resulting from measurement depends upon the number of decimal places; that is, a larger number of decimal places means a smaller probable error. In 2.3 inches the probable error is 0.05 inch, since 2.3 actually lies somewhere between 2.25 and 2.35.

In 1.426 inches there is a much smaller probable error of 0.0005 inch. If we add 2.300 + 1.426 and get an answer in thousandths, the answer, 3.726 inches, would appear to be precise to thousandths; but this is not true since there was a probable error of.05 in one of the addends. Also 2.300 appears to be precise to thousandths but in this example it is precise only to tenths.

It is evident that the precision of a sum is no greater than the precision of the least precise addend. It can also be shown that the precision of a difference is no greater than the less precise number compared. To add or subtract numbers of different orders, all numbers should first be rounded off to the order of the least precise number. In the foregoing example, 1.426 should be rounded to tenths-that is, 1.4.

This rule also applies to repeating decimals. Since it is possible to round off a repeating decimal at any desired point, the degree of precision desired should be determined and all repeating decimals to be added should be rounded to this level. Thus, to add the decimals generated by 1/3, 2/3, and 5/12 correct to thousandths, first round off each decimal to thousandths, and then add, as follows:

$$\begin{array}{r} .333 \\ .667 \\ .417 \\ \hline 1.417 \end{array}$$

When a common fraction is used in recording the results of measurement, the denominator of the fraction indicates the degree of precision. For example, a ruler marked in sixty-fourths of an inch has smaller divisions than one marked in sixteenths of an inch.

Therefore a measurement of 3 4/64 inches is more precise than a measure of 3 1/16 inches, even though the 16 two fractions are numerically equal. Remember that a measurement of 3 4/64 inches contains a probable error of only one-half of one sixty-fourth of an inch. On the other hand, if the smallest division on the ruler is one-sixteenth of an inch, then a measurement of 3 1/16 inches contains a probable error of one thirty-second of an inch.

ACCURACY

Even though a number may be very precise, which indicates that it was measured with an instrument having closely spaced divisions, it may not be very accurate. The accuracy of a measurement depends upon the relative size of the probable error when compared with the quantity being measured.

For example, a distance of 25 yards on a pistol range may be measured carefully enough to be correct to the nearest inch. Since there are 900 inches in 25 yards, this measurement is between 899.5 inches and 900.5 inches. When compared with the total of 900 inches, the 0.5-inch probable error is not very great.

On the other hand, a length of pipe may be measured rather precisely and found to be 3.2 inches long. The probable error here is 0.05 inch, and this measurement is thus more precise than that of the pistol range mentioned before. To compare the accuracy of the two measurements, we note that 0.05 inch out of a total of 3.2 inches is the same as 0.5 inch out of 32 inches. Comparing this with the figure obtained in the other example (0.5 inch out of 900), we conclude that the more precise measurement is actually the less accurate of the two measurements considered.

It is important to realize that the location of the decimal point has no bearing on the accuracy of the number. For example, 1.25 dollars represents exactly the same amount of money as 125 cents.

These are equally accurate ways of representing the same quantity, despite the fact that the decimal point is placed differently. Practice problems. In each of the following problems, determine which number of each pair is more accurate and which is more precise:

1. 3.72 inches or 2,417 feet
2. 2.5 inches or 17.5 inches
3. 5 3/4: inches or 12 7/8: inches
4. 34.2 seconds or 13 seconds

Answers:

1. 3.72 inches is more precise.
   2,417 feet is more accurate.
2. The numbers are equally precise.
   17.5 inches is more accurate.
3. 12 7/8; inches is more precise and more accurate.
4. 34.2 seconds is more precise and more accurate.

Percent of Error

The accuracy of a measurement is determined by the Relative error. The relative error is the ratio between the probable error and the quantity being measured.

This ratio is simply the fraction formed by using the probable error as the numerator and the measurement itself as the denominator. For example, suppose that a metal plate is found to be 5.4 inches long, correct to the nearest tenth of an inch. The maximum probable error is five hundredths of an inch (one-half of one tenth of an inch) and the relative error is found as follows:

$$\frac{\text{Probable error}}{\text{Measured value}} = \frac{0.05}{5.4} = \frac{5}{540}$$

Thus the relative error is 5 parts out of 540. Relative error is usually expressed as PERCENT OF ERROR, When the denominator of the fraction expressing the error ratio is divided into the numerator, a decimal is obtained. This decimal, converted to percent, gives the percent of error. For example, the error in the foregoing problem could be stated as 0.93 percent, since the ratio 5/540 reduces to 0.0093 (rounded off) in decimal form.

Significant Digits

The accuracy of a measurement is often described in terms of the number of significant digits used in expressing it. If the digits of a number resulting from measurement are examined one by one, beginning with the left-hand digit, the first digit that is not 0 is the first significant digit. For example, 2345 has four significant digits and 0.023 has only two significant digits.

The digits 2 and 3 in a measurement such as 0.023 inch signify how many thousandths of an inch comprise the measurement. The O's are of no significance in specifying the number of thousandths in the measurement; their presence is required only as "place holders" in placing the decimal point.

A rule that is often used states that the significant digits in a number begin with the first nonzero digit (counting from left to right) and end with the last digit. This implies that 0 can be a significant digit if it is not the first digit in the number. For example, 0.205 inch is a measurement having three significant digits. The 0 between the 2 and the 5 is significant because it is a part of the number specifying how many hundredths are in the measurement.

The rule stated in the foregoing paragraph fails to classify final 0's on the right. For example, in a number such as 4,700, the number of significant digits might be two, three, or four. If the 0's merely locate the decimal point (that is, if they show the number to be approximately forty-seven hundred rather than forty seven), then the number of significant digits is two.

However, if the number 4,700 represents a number such as 4,730 rounded off to the nearest hundred, there are three significant digits. The last 0 merely locates the decimal point. If the number 4,700 represents a number such as 4,700.4 rounded off, then the number of significant digits is four.

Unless we know how a particular number was measured, it is sometimes impossible to determine whether right-hand O's are the result of rounding off. However, in a practical situation it is normally possible to obtain information concerning the instruments used and the degree of precision of the original data before any rounding was done.

In a number such as 49.30 inches, it is reasonable to assume that the 0 in the hundredths place would not have been recorded at all if it were not significant. In other words, the instrument used for the measurement can be read to the nearest hundredth of an inch. The 0 on the right is thus significant.

This conclusion can be reached another way by observing that the 0 in 49.30 is not needed as a place holder in placing the decimal point. Therefore its presence must have some other significance. The facts concerning significant digits may be summarized as follows:

1. Digits other than 0 are always significant.
2. Zero is significant when it falls between significant digits.
3. Any final 0 to the right of the decimal point is significant.
4. When a 0 is present only as a place holder for locating the decimal point, it is not significant.
5. The following categories comprise the significant digits of any measurement number:
   a. The first nonzero left-hand digit is significant.
   b. The digit which indicates the precision of the

number is significant. This is the digit farthest to the right, except when the right-hand digit is 0. If it is 0, it may be only a place holder when the number is an integer.

c. All digits between significant digits are significant.

Practice problems: Determine the percent of error and the number of significant digits in each of the following measurements:

1. 5.4 feet
2. 0.00042 inch
3. 4.17 set
4. 147.50 miles

Answers:

1. Percent of error: 0.93%

   Significant digits: 2
2. Percent of error: 1.19%

   Significant digits: 2
3. Percent of error: 0.12%

   Significant digits: 3
4. Percent of error: 0.0034%

   Significant digits: 5

CALCULATING WITH APPROXIMATE NUMBERS

The concepts of precision and accuracy form the basis for the rules which govern calculation with approximate numbers (numbers resulting from measurement).

Addition and Subtraction

A sum or difference can never be more precise than the least precise number in the calculation. Therefore, before adding or subtracting approximate numbers, they should be rounded to the same degree of precision. The more precise numbers are all rounded to the precision of the least precise number in the group to be combined. For example, the numbers 2.95, 32.7, and 1.414 would be rounded to tenths before adding as follows:

$$\begin{array}{r} 3.0 \\ 32.7 \\ 1.4 \\ \hline \end{array}$$

Multiplication and Division

When two numbers are multiplied, the result often has several more digits than either of the original factors. Division also frequently produces more digits in the quotient than the original data possessed, if the division is "carried out" to several decimal places.

Results such as these appear to have more significant digits than the original measurements from which they came, giving the false impression of greater accuracy than is justified. In order to correct this situation, the following rule is used: In order to multiply or divide two approximate numbers having an equal number of significant digits, round the answer to the same number of significant digits as are shown in one of the original numbers.

If one of the original factors has more significant digits than the other, round the more accurate number before multiplying.

It should be rounded to one more significant digit than appears in the less accurate number; the extra digit protects the answer from the effects of multiple rounding. After performing the' multiplication or division, round the result to the same number of significant digits as are shown in the less accurate of the original factors.

Practice Problems:

1. Find the sum of the sides of a triangle in which the lengths of the three sides are as follows: 2.5 inches, 3.72 inches, and 4.996 inches.
2. Find the product of the length and width of a rectangle which is 2.95 feet long and 0.9046 foot wide.

Answers:

1. 11.2 inches
2. 2.67 square feet

MICROMETERS AND VERNIERS

Closely associated with the study of decimals is a measuring instrument known as a micrometer. The ordinary micrometer is capable of measuring accurately to one - thousandth of an inch. One-thousandth of an inch is about the thickness of a human hair or a thin sheet of paper. The parts of a micrometer are shown in figure 6-l.

MICROMETER SCALES

The spindle and the thimble move together. The end of the spindle (hidden from view in figure 6-l) is a screw with 40 threads per inch. Consequently, one complete turn of the thimble moves the spindle one-fortieth of an inch or 0.025 inch since 1/40 is equal to 0.025. The sleeve has 40 markings to the inch.

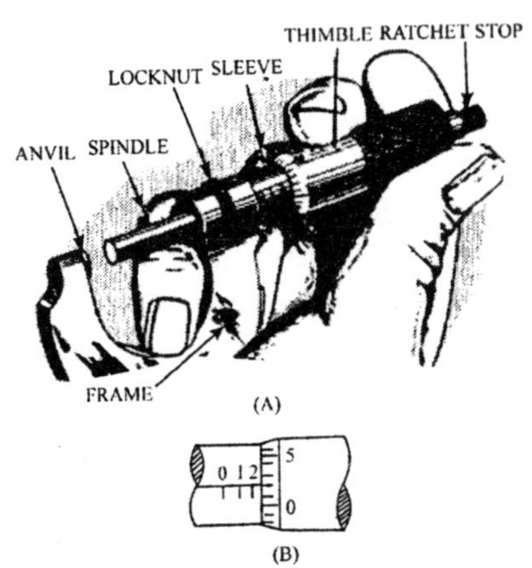

Fig. (A) Parts of a Micrometer; (B) Micrometer Scales

Thus each space between the markings on the sleeve is also 0.025 inch. Since 4 such spaces are 0.1 inch (that is, 4 × 0.025), every fourth mark is labeled in tenths of an inch for convenience in reading. Thus, 4 marks equal 0.1 inch, 8 marks equal 0.2 inch, 12 marks equal 0.3 inch, etc.

To enable measurement of a partial turn, the beveled edge of the thimble is divided into 25 equal parts.

Thus each marking on the thimble is 1/25 of a complete turn, or 1/25 of 1/40 of an inch. Multiplying 1/25 times 0.025 inch, we find that each marking on the thimble represents 0.001 inch.

READING THE MICROMETER

It is sometimes convenient when learning to read a micrometer to write down the component parts of the measurement as read on the scales and then to add them. For example, in figure 6-1 (B) there are two major divisions visible (0.2 inch).

One minor division is showing clearly (0.025 inch). The marking on the thimble nearest the horizontal or index line of the sleeve is the second marking (0.002 inch). Adding these parts, we have

$$\begin{array}{r} 0.200 \\ 0.025 \\ \underline{0.002} \\ 0.227 \end{array}$$

Thus, the reading is 0.227 inch. As explained previously, this is read verbally as "two hundred twenty-seven thousandths."

A more skillful method of reading the scales is to read all digits as thousandths directly and to do any adding mentally. Thus, we read the major division on the scale as "two hundred thousandths" and the minor division is added on mentally. The mental process for the above setting then would be "two hundred twenty-five; two hundred twenty-seven thousandths."

Practice Problems:

1. Read each of the micrometer settings shown in figure 6-2.

Answers:

(A) 0.750

(B) 0.201

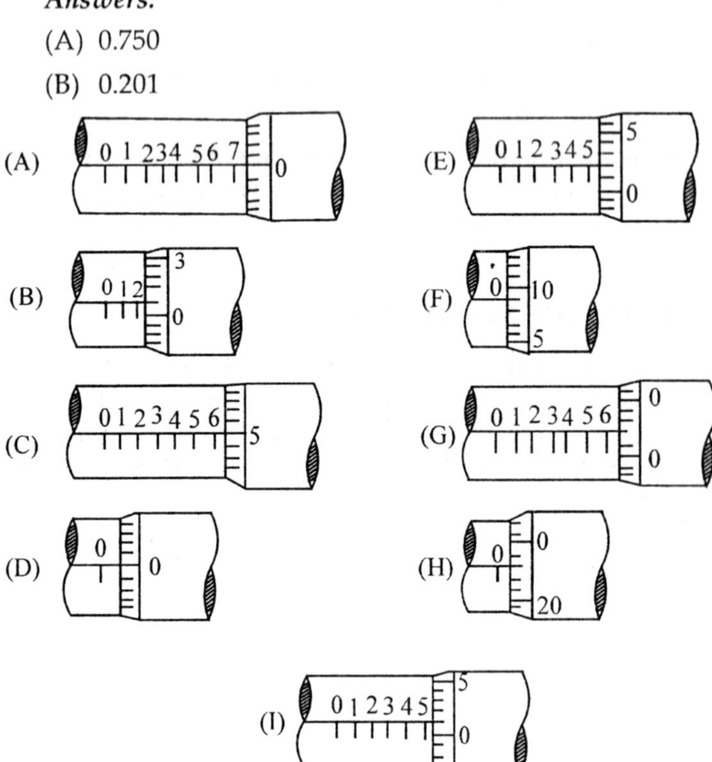

Fig. Micrometer settings

(C) 0.655

(D) 0.075

(E) 0.527

(F) 0.009

(G) 0.662

(H) 0.048

(I) 0.526

VENIRE

Sometimes the marking on the thimble of the micrometer does not fall directly on the index line of the sleeve. To make possible readings even smaller than thousandths, an ingenious device is introduced in the form of an additional scale. This scale, called a Vernier, was named after its inventor, Pierre Venire.

The Venire makes possible accurate readings to the ten-thousandth of an inch.

Principle of the Venire

Suppose a ruler has markings every tenth of an inch but it is desired to read accurately to hundredths. A separate scale (fig. 6-3) is added to the ruler. It has 10 markings on it that take up the same distance as 9 markings on the ruler scale. Thus, each space on the vernier is 1/10 of 9/10 inch, or 9/100 inch.

How much smaller is a space on the venire than a space on the ruler? The ruler space is 1/10 inch, or 10/100 and the venire space is 9/100 inch. The vernier space is smaller by the difference between these two numbers, as follows:

$$\frac{10}{100} - \frac{9}{100} = \frac{1}{100}$$

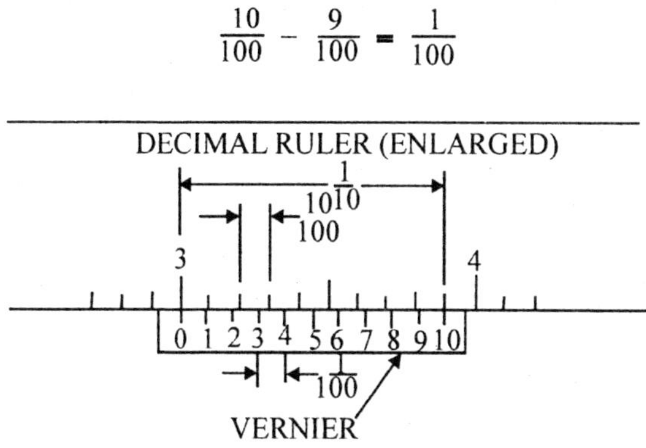

Fig. Vernier Scale

Each venire space is 1/100 inch smaller than a ruler space. As an example of the use of the venire scale, suppose that we are measuring the steel bar shown in figure. The end of the bar almost reaches the 3-inch mark on the ruler, and we estimate that it is about halfway between 2.9 inches and 3.0

inches. The vernier marks help us to decide whether the exact measurement is 2.94 inches, 2.95 inches, or 2.96 inches.

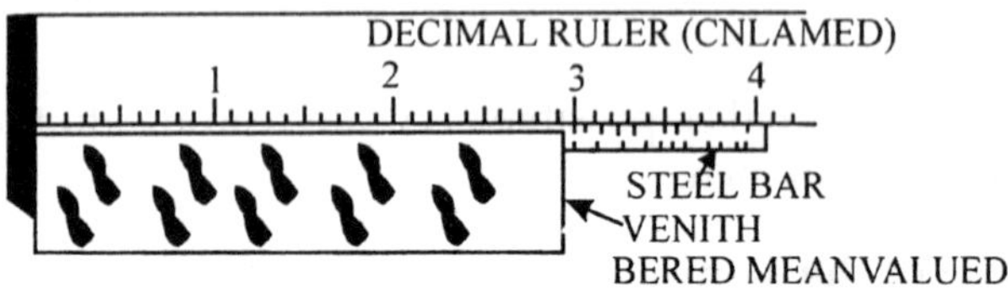

Fig. Measuring with a Vernier

The 0 on the vernier scale is spaced the distance of exactly one ruler mark (in this case, one tenth- of an inch) from the left hand end of the vernier. Therefore the 0 is at a position between ruler marks which is comparable to the position of the end of the bar. In other words, the 0 on the vernier is about halfway between two adjacent marks on the ruler, just as the end of the bar is about halfway between two adjacent marks. The 1 on the vernier scale is a little closer to alignment with an adjacent ruler mark; in fact, it is one hundredth of an inch closer to alignment than the 0. This is because each space on the vernier is one hundredth of an inch shorter than each space on the ruler.

Each successive mark on the vernier scale is one hundredth of an inch closer to 0 on the vernier must be five hundredths of an inch from the nearest ruler mark, since five increments, each one hundredth of an inch in size, were used before a mark was found in alignment.

We conclude that the end of the bar is five hundredths of an inch from the 2.9 mark on the ruler, since its position between marks is exactly comparable to that of the 0 on the vernier scale. Thus the value of our measurement is 2.95 inches.

The foregoing example could be followed through for any distance between markings. Suppose the 0 mark fell seven tenths of the distance between ruler markings. It would take seven vernier markings, a loss of one-hundredth of an inch each time, to bring the marks in line at 7 on the vernier.

The vernier principle may be used to get fine linear readings, angular readings, etc. The principle is always the same. The vernier has one more marking than the number of markings on an equal space of the conventional scale of the measuring instrument. For example, the vernier caliper has 25 markings on the vernier for 24 on the caliper scale. The caliper is marked off to read to fortieths (0.025) of an inch, and the vernier extends the accuracy to a thousandth of an inch.

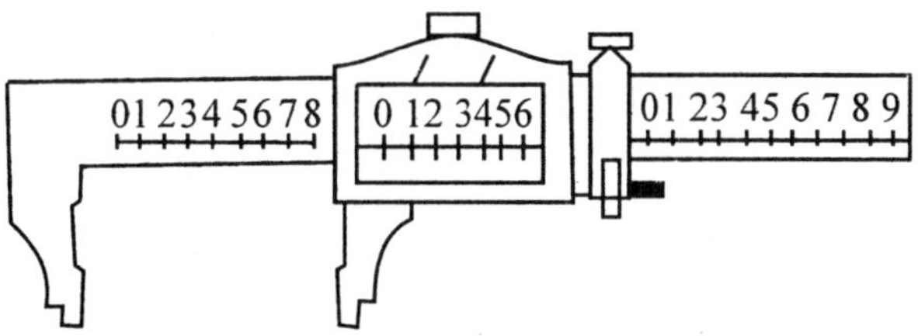

Fig. A Vernier Caliper

Vernier Micrometer

By adding a vernier to the micrometer, it is possible to read accurately to one ten-thousandth of an inch. The vernier markings are on the sleeve of the micrometer and are parallel to the thimble markings. There are 10 divisions on the vernier that occupy the same space as 9 divisions on the thimble. Since a thimble space is one thousandth of an inch, a vernier space is 1/10 of 9/1000 inch, or 9/10000 inch. It is 1/10000 inch less than a thimble space. Thus, as in the preceding explanation of verniers, it is possible to read the nearest ten-thousandth of an inch by reading the vernier digit whose marking coincides with a thimble marking.

In figure, the last major division showing fully on the sleeve index is 3. The third minor division is the last mark clearly showing (0.075). The thimble division nearest and below the index is the 8 (0.008). The vernier marking that matches a thimble marking is the fourth (0.0004). Adding them all together, we have,

$$\begin{array}{r} 0.3000 \\ 0.0750 \\ 0.0080 \\ \underline{0.0004} \\ 0.3834 \end{array}$$

The reading is 0.3834 inch. With practice these readings can be made directly from the micrometer, without writing the partial readings.

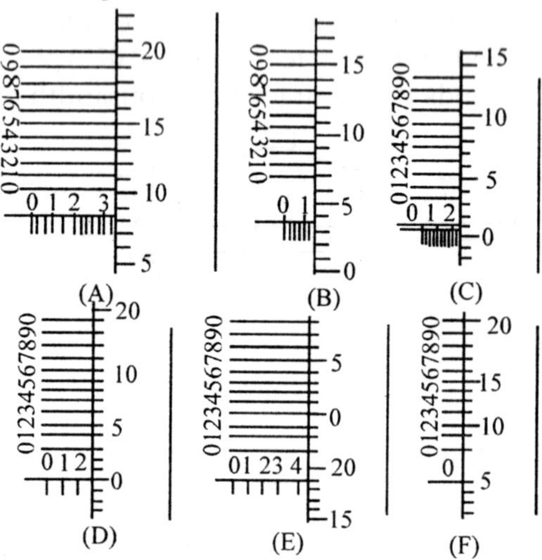

Fig. Vernier Micrometer Settings

Practice problems:

1. Read the micrometer settings in figure 6-6.

Answers:

1. (A) See the foregoing example.
   (B) 0.1539
   (C) 0.2507
   (D) 0.2500
   (E) 4.4690
   (F) 0.0552

Chapter 3

Exponents and Radicals

The operation of raising a number to a power is a special case of multiplication in which the factors are all equal. In examples such as $4^2 = 4 \times 4 = 16$ and $5^3 = 5 \times 5 \times 5 = 125$, the number 16 is the second power of 4 and the number 125 is the third power of 5. The expression 5^3 means that three 5's are to be multiplied together. Similarly, 4^2 means 4×4. The first power of any number is the number itself. The power is the number of times the number itself is to be taken as a factor.

The process of finding a root is the inverse of raising a number to a power. A root is a special factor of a number, such as 4 in the expression $4^2 = 16$. When a number is taken as a factor two times, as in the expression $4 \times 4 = 16$, it is called a square root. Thus, 4 is a square root of 16. By the same reasoning, 2 is a cube root of 8. since $2 \times 2 \times 2$ is equal to 8. This relationship is usually written as $2^3 = 8$.

POWERS AND ROOTS

A power of a number is indicated by an EXPONENT, which is a number in small print placed to the right and toward the top of the number. Thus, in $4^3 = 64$, the number 3 is the EXPONENT of the number 4. The exponent 3 indicates that the number 4, called the BASE, is to be raised to its third power. The expression is read "4 to the third power (or 4 cubed) equals 64." Similarly, $5^2 = 25$ is read "5 to the second power (or 5 squared) equals 25." Higher powers are read according to the degree indicated; for example, "fourth power," "fifth power," etc. When an exponent occurs, it must always be written unless its value is 1. The exponent 1 usually is not written, but is understood. For example, the number 5 is actually 5. When we work with exponents, it is important to remember that any number that has no written exponent really has an exponent equal to 1.

A root of a number can be indicated by placing a radical sign, $\sqrt{\ }$, over the number and showing the root by placing a small number within the notch of the radical sign. Thus, $\sqrt[3]{64}$ indicates the cube root of 64, and $\sqrt[3]{32}$ indicates the fifth root of 32. The number that indicates the root is called the Index of the root. In the case of the square root, the index, 2, usually is not shown. When a radical has no index, the square root is understood to be the one desired.

For example, $\sqrt{36}$ indicates the square root of 36. The line above the number whose root is to be found is a symbol of grouping called the vinculum. When the radical symbol is used, a vinculum, long enough to extend over the entire expression whose root is to be found, should be attached.

Practice Problems: Raise to the indicated power or find the root indicated.

Answers:

1. 2^3
2. 6^3
3. 4^3
4. 25^3
5. $\sqrt{16}$
6. $\sqrt[3]{8}$
7. $\sqrt[3]{125}$
8. $\sqrt[5]{32}$

Answers:

1. 8
2. 36
3. 64
4. 15.6255.4
5. 4
6. 2
7. 5

NEGATIVE INTEGERS

Raising to a power is multiplication in which all the numbers being multiplied together are equal. The sign of the product is determined, as in ordinary multiplication, by the number of minus signs. The number of minus signs is odd or even, depending on whether the exponent of the base is odd or even. For example, in the problem

$$2^3 = (-2)(-2)(-2) = -8$$

There are three minus signs. The result is negative. In

$$(-2)^6 = 64$$

There are six minus signs. The result is positive.

Thus, when the exponent of a negative number is odd, the power is negative; when the exponent is even, the power is positive. As other examples, consider the following:

$$(-3)^4 = 81$$

$$\left(-\frac{2}{5}\right)^3 = -\frac{8}{125}$$

$$(-2)^3 = 225$$

$$(-1)^5 = -1$$

Positive and negative numbers belong to the class called REAL NUMBERS. The square of a real number is positive. For example, $(-7)^2 = 49$ and $7^2 = 49$. The expression $(-7)^2$ is read "minus seven squared." Note that either seven squared or minus seven squared gives us +49. We cannot obtain –49 or any other negative number by squaring any real number, positive or negative.

Since there is no real number whose square is a negative number, it is sometimes said that the square root of a negative number does not exist. However, an expression under a square root sign may take on negative values. While the square root of a negative number cannot actually be found, it can be indicated. The indicated square root of a negative number is called an Imaginary Number. The number, for example, is said to be imaginary. It is read "square root of minus seven."

FRACTIONS

We recall that the exponent of a number tells the number of times that the number is to be taken as a factor. A fraction is raised to a power by raising the numerator and the denominator separately to the power indicated. The expression $\left(\frac{3}{7}\right)^2$ means 3/7 is used twice as a factor. Thus,

$$\left(\frac{3}{7}\right)^2 = \frac{3}{7} \times \frac{3}{7} = \frac{3^2}{7^2} = \frac{9}{49}$$

Similarly,

$$\left(-\frac{1}{5}\right)^2 = \frac{1}{25}$$

Since a minus sign can occupy any one of three locations in a fraction, notice that evaluating $(-1/5)^2$ is equivalent to

$$(-1)^2\left(\frac{1}{5}\right)^2 \text{ or } \frac{(-1)^2}{5^2} \text{ or } \frac{-1^2}{(-5)^2}$$

The process of taking a root of a number is the inverse of the process of raising the number to a power, and the method of taking the root of a fraction is similar. We may simply take the root of each term separately and write the result as a fraction.

Consider the following examples:

1. $\sqrt{\frac{36}{49}} = \sqrt{\frac{36}{\sqrt{49}}} = \frac{6}{7}$

2. $-\sqrt[3]{\frac{8}{49}} = \frac{\sqrt[3]{36}}{\sqrt[3]{125}} = \frac{2}{3}$

Practice Problems: Find the values for. the indicated operations:

1. $\left(\frac{1}{3}\right)^2$

2. $\left(\frac{3}{4}\right)^2$

3. $\left(\frac{6}{5}\right)^2$

4. $\left(\frac{2}{3}\right)^3$

5. $\sqrt{\frac{16}{36}}$

6. $\sqrt{\frac{16}{25}}$

7. $\sqrt[3]{\frac{8}{27}}$

8. $\sqrt{\frac{9}{49}}$

Answers:

1. 1/9
2. 9/16
3. 36/25
4. 8/27
5. 4/6

6. 4/5
7. 2/3
8. 3/7

DECIMALS

When a decimal is raised to a power, the number of decimal places in the result is equal to the number of places in the decimal multiplied by the exponent. For example, consider (0.12). There are two decimal places in 0.12 and 3 is the exponent. Therefore, the number of places in the power will be 3(2) = 6. The result is as follows:

$$(0.12)^3 = 0.001728$$

The truth of this rule is evident when we recall the rule for multiplying decimals. Part of the rule states: Mark off as many decimal places in the product as there are decimal places in the factors together. If we carry out the multiplication, (0.12) × (0.12) × (0.12), it is obvious that there are six decimal places in the three factors together. The rule can be shown for any decimal raised to any power by simply carrying out the multiplication indicated by the exponent.

Consider these examples:

$$(1.4)^2 = 1.96$$

$$(0.12)^2 = 0.0144$$

$$(0.4)^3 = 0.064$$

$$(0.02)^2 = 0.0004$$

$$(0.2)^2 = 0.04$$

Finding a root of a number is the inverse of raising a number to a power. To determine the number of decimal places in the root of a perfect power, we divide the number of decimal places in the radicand by the index of the root. Notice that this is just the opposite of what was done in raising a number to a power.

Consider $\sqrt{0.0625}$. The square root of 625 is 25. There are four decimal places in the radicand, 0.0625, and the index of the root is 2. Therefore, 4 ÷ 2 = 2 is the number of decimal places in the root. We have

$$\sqrt{0.0625} = 0.25$$

Similarly,

$$\sqrt{1.69} = 1.3$$

$$\sqrt[3]{0.027} = 0.3$$

$$\sqrt[3]{1.728} = 1.2$$

$$\sqrt[4]{0.0001} = 0.1$$

LAWS OF EXPONENTS

All of the laws of exponents may be developed directly from the definition of exponents. Separate laws are stated for the following five cases:

1. Multiplication.
2. Division.
3. Power of a power.
4. Power of a product.
5. Power of quotient.

MULTIPLICATION

To illustrate the law of multiplication, we examine the following problem:

$$4^3 \times 4^2 = ?$$

Recalling that 4^3 means $4 \times 4 \times 4$ and 4^2 means 4×4, we see that 4 is used as a factor five times. Therefore $4^3 \times 4^2$ is the same as 4^5. This result could be written as follows:

$$4^3 \times 4^2 = 4 \times 4 \times 4 \times 4 \times 4 = 4^5$$

Notice that three of the five 4's came from the expression 4^3, and the other two 4's came from the expression 4^2. Thus we may rewrite the problem as follows:

$$4^3 \times 4^2 = 4 \times 4 \times 4 \times 4 \times 4 = 4^5$$

The law of exponents for multiplication may be stated as follows: To multiply two or more powers having the same base, add the exponents and raise the common base to the sum of the exponents. This law is further illustrated by the following examples:

$$2^3 \times 2^4 = 2^7$$

$$3 \times 3 = 3^3$$

$$15^4 \times 15^2 = 15^6$$

$$10^2 \times 10^{0.5} = 15^{2.5}$$

Common Errors

It is important to realize that the base must be the same for each factor, in order to apply the laws of exponents. For example, $2^3 \times 3^2$ is neither 2^5 nor 3^5.

There is no way to apply the law of exponents to a problem of this kind. Another common mistake is to multiply the bases together. For example, this kind of error in the foregoing problem would imply that $2^3 \times 3^2$ is equivalent to 6^5, or 7776. The error of this may-be proved as follows:

$$2^3 \times 3^2 = 8 \times 9 = 72$$

DIVISION

The law of exponents for division may be developed from the following example:

$$6^7 + 6^5 = \frac{\not{6} \times \not{6} \times \not{6} \times \not{6} \times \not{6} \times 6 \times 6}{\not{6} \times \not{6} \times \not{6} \times 6 \times 6} = 6^2$$

Cancellation of the five 6's in the divisor with five of the 6's in the dividend leaves only two 6's, the product of which is 6^2.

This result can be reached directly by noting that 6^2 is equivalent to $6^{(7-5)}$. In other words, we have the following:

$$0 = 6^2$$

Therefore the law of exponents for division is as follows: To divide one power into another having the same base, subtract the exponent of the divisor from the exponent of the dividend. Use the number resulting from this subtraction as the exponent of the base in the quotient. Use of this rule sometimes produces a negative exponent or an exponent whose value is 0.

POWER OF A POWER

Consider the example $(3^2)^4$. Remembering that an exponent shows the number of times the base is to be taken as a factor and noting in this case that 3^2 is considered the base, we have

$$(3^2)^4 = 3^2 \times 3^2 \times 3^2 \times 3^2$$

Also in multiplication we add exponents. Thus,

$$3^2 \times 3^2 \times 3^2 \times 3^2 = 3^{(2+2+2+2)} = 38$$

Therefore,

$$(3^2)^4 = 3^{(4\times2)} = 3^8$$

The laws of exponents for the power of a power may be stated as follows: To find the power of a power, multiply the exponents. It should be noted that this case is the only one in which multiplication of exponents is performed.

POWER OF A PRODUCT

Consider the example $(3 \times 2 \times 5)^3$. We know that

$$(3 \times 2 \times 5)^3 = (3 \times 2 \times 5)(3 \times 2 \times 5)(3 \times 2 \times 5)$$

Thus 3, 2, and 5 appear three times each as factors, and we can show this with exponents as 3^3, 2^3, and 5^3. Therefore,

$$(3 \times 2 \times 5)^3 = 3^2 \times 2^3 \times 5^3$$

The law of exponents for the power of a product is as follows: The power of a product is equal to the product obtained when each of the original factors is raised to the indicated power and the resulting powers are multiplied together.

POWER OF A QUOTIENT

The law of exponents for a power of an indicated quotient may be developed from the following example:

$$\left(\frac{2}{3}\right)^3 = \frac{2}{3}.\frac{2}{3}.\frac{2}{3} = \frac{2.2.2}{3.3.3} = \frac{2^3}{3^3}$$

Therefore,

$$\left(\frac{2}{3}\right)^3 = \frac{2^3}{3^3}$$

The law is stated as follows: The power of a quotient is equal to the quotient obtained when the dividend and divisor are each raised to the indicated power separately, before the division is performed.

Practice problems: Raise each of the following expressions to the indicated power:

1. $(3^2 . 2^3)^2$
2. $35 \div 3^2$
3. $\left(\frac{3.2}{5.6}\right)^3$
4. $(-3^2)^3$
5. $\frac{5^3}{5}$
6. $(3.2.7)^2$

Answers:

1. $3^4 \times 2^6 = 5{,}184$
2. 27
3. 1/125
4. $[(-3)^2]^3 = 729$
5. 25
6. $9 \times 4 \times 49 = 1{,}764$

SPECIAL EXPONENTS

Thus far in this discussion of exponents, the emphasis has been on exponents which are positive integers. There are two types of exponents which are not positive integers, and two which are treated as special cases even though they may be considered as positive integers.

ZERO AS AN EXPONENT

Zero occurs as an exponent in the answer to a problem such as $4^3 + 4^3$. The law of exponents for division states that the exponents are to be subtracted. This is illustrated as follows:

$$\frac{4^3}{4^3} = 4^{(3-3)} = 4^0$$

Another way of expressing the result of dividing 4^3 by 4^3 is to use the fundamental axiom which states that any number divided by itself is 1. In order for the laws of exponents to hold true in all cases, this must also be true when any number raised to a power is divided by itself. Thus, $4^3/4^3$ must equal 1. Since $4^3/4^3$ has been shown to be equal to both 4^0 and 1, we are forced to the conclusion that $4^0 = 1$.

By the same reasoning,

$$\frac{5}{5} = 5^{1-1} = 5^0$$

Also,

$$5/5 = 1$$

Therefore,

$$5^0 = 1$$

Thus we see that any number divided by itself results in a 0 exponent and has a value of 1. By definition then, any number (other than zero) raised to the zero power equals 1. This is further illustrated in the following examples:

ONE AS AN EXPONENT

The number 1 arises as an exponent sometimes as a result of division. In the example $5^3/5^2$ we subtract the exponents to get

$$5^{3-2} = 5^1$$

This problem may be worked another way as follows:

$$\frac{5^3}{5^2} = \frac{\not{5}.\not{5}.5}{\not{5}.\not{5}} = 5$$

Therefore,

$$5^1 = 5$$

We conclude that any number raised to the first power is the number itself. The exponent 1 usually is not written but is understood to exist.

NEGATIVE EXPONENTS

If the law of exponents for division is extended to include cases where the exponent of the denominator is larger, negative exponents arise. Thus,

$$\frac{3^2}{3^5} = 3^{2-5} = 3^{-3}$$

Another way of expressing this problem Is as follows:

$$\frac{3^2}{3^5} = \frac{\not{3}.\not{3}}{\not{3}.\not{3}.3.3.3} = \frac{1}{3^3}$$

Therefore,

$$3^{-3} = \frac{1}{3^3}$$

We conclude that a number N with a negative exponent is equivalent to a fraction having the following form: Its numerator is 1; its denominator is N with a positive exponent whose absolute value is the same as the absolute value of the original exponent. In symbols, this rule may be stated as follows:

$$N^{-a} = \frac{1}{N^a}$$

Also,

$$\frac{1}{N^{-a}} = N^a$$

The following examples further illustrate the rule:

$$5^{-1} = \frac{1}{5}$$

$$6^{-2} = \frac{1}{6^2}$$

$$4^{-12} = \frac{1}{4^{12}}$$

$$\frac{1}{3^{-2}} = 3^2$$

Notice that the sign of an exponent may be changed by merely moving the expression which contains the exponent to the other position in the fraction. The sign of the exponent is changed as this move is made. For example,

$$\frac{1}{10^{-2}} = 1 \div \frac{1}{10^2} = 1 \times \frac{10^2}{1}$$

Therefore,

$$\frac{1}{10^{-2}} = \frac{10^2}{1} 1 \times \frac{10^2}{1}$$

By using the foregoing relationship, a problem such as $3+5^{-4}$ may be simplified as follows:

$$\frac{3}{5^{-4}} = 3 \times \frac{1}{5^{-4}}$$

$$3 \times \frac{5^4}{1}$$

$$3 \times 5^4$$

FRACTIONAL EXPONENTS

Fractional exponents obey the same laws as do integral exponents. For example,

$$4^{1/2} \times 4^{1/2} = 4^{(1/2+1/2)} = 4^{2/2}$$

$$4^1 = 4$$

Another way of expressing this would be

$$4^{1/2} \times 4^{1/2} = 4^{(1/2+1/2)} = 4^{(1/2)^2} = 4^{(1/2\times 2)}$$

$$4^1 = 4$$

Observe that the number $4^{1/2}$, when squared in the foregoing example, produced the number 4 as an answer. Recalling that a square root of a number N is a number x such that $x^2 = N$, we conclude that $4^{1/2}$ is equivalent to a. Thus we have a definition, as follows: A fractional exponent of the form $1/r$ indicates a root, the index of which is r. This is further illustrated in the following examples:

$$2^{1/2} = \sqrt{2}$$

$$4^{1/3} = \sqrt[3]{4}$$

$$6^{2/3} = (6^{1/3}) = (\sqrt[3]{6})^2$$

Also,

$$6^{2/3} = (6^2)^{1/3} = \sqrt[3]{326}$$

Notice that in an expression such as $8^{2/3}$ we can either find the cube root of 8 first or square 8 first, as shown by the following example:

$$(8^{1/3})^2 = 2^2 \text{ and } (8^2)^{1/3} = \sqrt[3]{64} = 4$$

All the numbers in the evaluation of $8^{2/3}$ remain small if the cube root is found before raising the number to the second power. This order of operation is particularly desirable in evaluating a number like $64^{5/6}$, If 64 were first raised to the fifth power, a large number would result.

It would require a great deal of unnecessary effort to find the sixth root of 64^5. The result is obtained easily, if we write

$$64^{5/6} = (64^{1/6})^5 = 2^5 = 32$$

If an improper fraction occurs in an exponent, such as 7/3 in the expression $2^{7/3}$, it is customary to keep the fraction in that form rather than express it as a mixed number. In fraction form an exponent shows immediately what power is intended and what root is intended. However, $2^{7/3}$ can be expressed in another form and simplified by changing the improper fraction to a mixed number and writing the fractional part in the radical form as follows:

$$2^{7/3} = 2^{2+1/3} = 2^5 = 2^{1/3} = 4\sqrt[3]{2}$$

The law of exponents for multiplication may be combined with the rule for fractional exponents to solve problems of the following type:

Problem: Evaluate the expression $4^{2.5}$

Solution:

$$4^{2.5} = 4^2 \times 4^{0.5}$$
$$= 16 \times 4^{1/2}$$
$$= 16 \times 2 = 32$$

Practice problems:

1. Perform the indicated division: $\frac{2}{2^{1/3}}$
2. Find the product: $7^{2/5} \times 7^{1/10} \times 7^{3/10}$
3. Rewrite with a positive exponent and simplify: $9^{-1/2}$
4. Evaluate $100^{3/2}$
5. Evaluate $(8^0)^5$

Answers:

1. $2^{3/3} + 2^{1/3} = \sqrt[3]{4}$
2. $7^{8/10}$
3. $\frac{1}{9^{1/2}} = \frac{1}{3}$
4. 1,000
5. 1

SCIENTIFIC NOTATION AND POWERS OF 10

Technicians, engineers, and others engaged in scientific work are often required to solve problems involving very large and very small numbers. Problems such as

$$\frac{22.684 \times 0.00189}{0.0713 \times 83 \times 7}$$

are not uncommon. Solving such problems by the rules of ordinary arithmetic is laborious and time consuming.

Moreover, the tedious arithmetic process lends itself to operational errors. Also there is difficulty in locating the decimal point in the result. These difficulties can be greatly reduced by a knowledge of the powers of 10 and their use.

The laws of exponents form the basis for calculation using powers of 10. The following list includes several decimals and whole numbers expressed as powers of 10:

$$10{,}000 = 10^4$$
$$1{,}000 = 10^3$$
$$100 = 10^2$$
$$10 = 10^1$$
$$0.1 = 10^{-2}$$
$$0.001 = 10^{-3}$$
$$0.0001 = 10^{-4}$$

The concept of scientific notation may be demonstrated as follows:

$$60{,}000 = 6.0000 \times 10.000 = 6 \times 10^4$$
$$538 = 5.38 \times 100 = 5.38 \times 10$$

Notice that the final expression in each of the foregoing examples involves a number between 1 and 10, multiplied by a power of 10. Furthermore, in each case the exponent of the power of 10 is a number equal to the number of digits between the new position of the decimal point and the original position (understood) of the decimal point.

We apply this reasoning to write any number in scientific notation; that is, as a number between 1 and 10 multiplied by the appropriate power of 10. The appropriate power of 10 is found by the following mechanical steps:

1. Shift the decimal point to standard position, which is the position immediately to the right of the first nonzero digit.
2. Count the number of digits between the new position of the decimal point and its original position. This number indicates the value of the exponent for the power of 10.

3. If the decimal point is shifted to the left, tie sign of the exponent of 10 is positive; if the decimal point is shifted to the right, the sign of the exponent is negative.

The validity of this rule, for those cases in which the exponent of 10 is negative, is demonstrated as follows:

$$0.00657 = 6057 \times 0.001$$
$$= 6.57 \times 10^{-3}$$
$$0.343 = 3.48 \times 0.1$$
$$= 3.48 \times 10^{-1}$$

Further examples of the use of scientific notation are given as follows:

Multiplication Using Powers of 10

From the law of exponents for multiplication we recall that to multiply two or more powers to the same base we add their exponents. Thus,

$$10^4 \times 10^2 = 10^6$$

We see that multiplying powers of 10 together is an application of the general rule. This is demonstrated in the following examples:

1. $10{,}000 \times 100 = 10^4 \times 10^2 = 10^4 + 7 = 10^6$
2. $0.0000001 \times 0.001 = 10^{-7} \times 10^{-3} = 10^{-7+(-3)} = 10^{-10}$
3. $10{,}000 \times 0.001 = 10^4 \times 10^{-3} = 10^{4-3} = 10$
4. $23{,}000 \times 500 = ?$

$23{,}000 = 2.3 \times 10^4$

$500 = 5 \times 10^2$

Therefore,

$$23{,}000 \times 500 = 2.3 \times 10^4 \times 5 \times 10^2$$
$$= 2.3 \times 5 \times 10^4 \times 10^2 = 11.5 \times 10^6$$
$$= 1.15 \times 10^7$$

5. $62{,}000 \times 0.003 \times 4{,}600 = ?$

$62{,}000 = 6.2 \times 10^4$

$0.0003 = 3 \times 10^{-4}$

$4{,}600 = 4.6 \times 10^3$

Therefore,

$$62{,}000 \times 0.0003 \times 4{,}600 = 6.2 \times 3 \times 4.6 \times 10^4 \times 10^{-4} \times 10^3$$
$$= 85.56 \times 10^3 = 8.556 \times 10^4$$

Practice problems: Multiply, using powers of 10. For the purposes of this exercise, treat all numbers as exact numbers:

1. $10{,}000 \times 0.001 \times 100$
2. $0.000350 \times 5{,}000{,}000 \times 0.0004$
3. $3{,}875 \times 0.000032 \times 3{,}000{,}000$
4. $7{,}000 \times 0.015 \times 1.78$

Answers:

1. 1.0×10^3
2. $.0 \times 10^{-1}$
3. 3.72×10^5
4. 1.869×10^2

Division Using Powers of 10

The rule of exponents for division states that, for powers of the same base, the exponent of the denominator is subtracted from the exponent of the numerator. Thus,

$$\frac{10^7}{10^2} = 10^{7-3} = 10^4$$

It should be remembered that power may be transferred from numerator to denominator or from denominator to numerator by simply changing the sign of the exponent. The following examples illustrate the use of this rule for powers of 10:

$$\frac{72{,}000}{0.0012} = \frac{7.2 \times 10^4}{1.2 \times 10^{-3}} = \frac{7.2}{1.2} \times 10^4 \times 10^3 = 6 \times 10^7$$

2. $$\frac{44 \times 10^{-4}}{11 \times 10^{-5}} = \frac{44}{11} \times 10^{-4} \times 10^5 = 4 \times 10$$

Combined Multiplication and Division

Using the rules already shown, multiplication and division involving powers of 10 may be combined.

The usual method of solving such problems is to multiply and divide alternately until the problem is completed. For example,

$$\frac{36{,}000 \times 1.1 \times 0.06}{0.012 \times 2{,}200}$$

Rewriting this problem in scientific notation, we have

$$\frac{3.6 \times 10^4 \times 1.1 \times 6 \times 10^{-2}}{1.2 \times 10^{-2} \times 2.2 \times 10^2} = \frac{3.6 \times 1.1 \times 6}{1.2 \times 2.2} \times 10 = 9 \times 10 = 90$$

Notice that the elimination of O's, wherever possible, simplifies the computation and makes it an easy matter to place the decimal point. SIGNIFICANT DIGITS.-One of the most important advantages of scientific notation is the fact that it simplifies the task of determining the number of significant digits in a number. For example, the fact that the number 0.00045 has two significant digits is sometimes obscured by the presence of the 0's. The confusion can be avoided by writing the number in scientific notation, as follows:

$$0.00045 = 4.5 \times 10^{-4}$$

Practice problems: Express the numbers in the following problems in scientific notation and round off before performing the calculation. In each problem, round off calculation numbers to one more digit than the number of significant digits in the least accurate number; round the answer to the number of significant digits in the least accurate number:

1. $\dfrac{0.000063 \times 50.4 \times 0.007213}{780 \times 0.682 \times 0.018}$

2. $\dfrac{0.015 \times 216 \times 1.78}{72 \times 0.0624 \times 0.0353}$

3. $\dfrac{0.000079 \times 0.00036}{29 \times 10}$

Answers:

1. 2.4×10^{-6}
2. 3.6×10
3. 9.8×10^{-2}

Other Applications

The applications of powers of 10 may be broadened to include problems involving reciprocals and powers of products.

Reciprocals. The following example illustrates the use of powers of 0 in the formation of a reciprocal:

$$\frac{1}{250,000 \times 300 \times 0.02} = \frac{1}{2.5 \times 10^5 \times 3 \times 10^2 \times 2 \times 10^{-2}}$$

$$= \frac{10^{-5}}{2.5 \times 3 \times 2} = \frac{10^{-5}}{15}$$

Rather than write the numerator as 0.00001, write it as the product of two factors, one of which may be easily divided, as follows:

$$\frac{10^{-5}}{15} = \frac{10^2 \times 10^{-7}}{15} = \frac{100}{15} \times 10^{-7} = 6.67 \times 10^{-7} = 0.000000667$$

Power of a Product: The following example illustrates the use of powers of 10 in finding the power of a product:

$$(80,000 \times 2 \times 10^5)^2 = (8 \times 10^4 \times 2 \times 10^3)^2$$

$$= 8^2 \times 2^2 \times (10^{4+5})^2$$

$$= 64 \times 4 \times 10^{18} = 256 \times 10^{18}$$

$$= 2.56 \times 10^{20}$$

RADICALS

An expression such as, $\sqrt{2}, \sqrt[3]{5}$ or $\sqrt{a+b}$ that exhibits a radical sign, is referred to as a RADICAL. We have already worked with radicals in the form of fractional exponents, but it is also frequently necessary to work with them in the radical form. The word "radical" is derived from the Latin word "radix," which means "root."

The word "radix" itself is more often used in modern mathematics to refer to the base of a number system, such as the base 2 in the binary system. However, the word "radical" is retained with its original meaning of "root."

The radical symbol appears to be a distortion of the initial letter "r" from the word "radix." With long usage, the r gradually lost its significance as a letter and became distorted into the symbol as we use it. The vinculum helps to specify exactly which of the letters and numbers following the radical sign actually belong to the radical expression.

The number under a radical sign is the Radicand. The index of the root (except in the case of a square root) appears in the trough of the radical sign. The index tells what root of the radicand is intended. For example, in $\sqrt[5]{32}$, the radicand is 32 and the index of the root is 5. The fifth root of 32 is intended. In $\sqrt{50}$, the square root of 50 is intended. When the index is 2, it is not written, but is understood.

If we can find one square root of a number we can always find two of them. Remember $(3)^2$ is 9 and $(-3)^2$ is also 9. Likewise $(4)^2$ and $(-4)^2$ both equal 16 and $(5)^2$ and $(-5)^2$ both equal 25. Conversely, $\sqrt{9}$ is +3 or –3, $\sqrt{16}$ is +4 or –4, and $\sqrt{25}$ is +5 or –5.

When we wish to show a number that may be either positive or negative, we may use the symbol + which is read "plus or minus." Thus ± 3 means "plus or minus 3." Usually when a number is placed under the radical sign, only its positive root is desired and, unless otherwise specified, it is the only root that need be found.

COMBINING RADICALS

A number written in front of another number and intended as a multiplier is called a Coefficient. The expression $5x$ means 5 times x; means a times y; and $7\sqrt{2}$ means 7 times $\sqrt{2}$. In these examples, 5 is the coefficient of x, a is the coefficient of y, and 7 is the coefficient of $\sqrt{2}$.

Radicals having the same index and the same radicand are SIMILAR. Similar radicals may have different coefficients in front of the radical sign. For example, $3\sqrt{2}$, $\sqrt{2}$, and $1/5\sqrt{2}$ are similar radicals. When a coefficient is not written, it is understood to be 1. Thus, the coefficient of $\sqrt{2}$ is 1. The rule for adding radicals is the same as that stated for adding denominate numbers: Add only units of the same kind. For example, we could add $2\sqrt{3}$ and $4\sqrt{3}$ because the "unit" in each of these numbers is the same ($\sqrt{3}$). By the same reasoning, we could not add $2\sqrt{3}$ and $4\sqrt{5}$ because these are not similar radicals.

Addition and Subtraction

When addition or subtraction of similar radicals is indicated, the radicals are combined by adding or subtracting their coefficients and placing the result in front of the radical. Adding 3 and 5 is similar to adding $3\sqrt{2}$ bolts and $5\sqrt{2}$ bolts. The following examples illustrate the addition and subtraction of similar radical expressions:

1. $3\sqrt{2}+5\sqrt{2}=8\sqrt{2}$
2. $1/2\left(\sqrt[4]{3}\right)+1/3\left(\sqrt[4]{3}\right)=5/6\left(\sqrt[4]{3}\right)$
3. $\sqrt{5}-6\sqrt{5}+2\sqrt{5}=-3\sqrt{5}$
4. $-5\sqrt[3]{7}-2\sqrt[3]{7}+7\sqrt[3]{7}=0$

Example 4 illustrates a case that is sometimes troublesome. The sum of the coefficients, –5, –2, and 7, is 0. Therefore, the coefficient of the answer would be 0, as follows:

$$0\left(\sqrt[3]{7}\right)=0\times\sqrt[3]{7}$$

Thus the final answer is 0, since 0 multiplied by any quantity is still 0.

Practice problems: Perform the indicated operations:

1. $4\sqrt{3}-\sqrt{3}+5\sqrt{3}$

2. $\frac{1}{2}\sqrt{6}+\sqrt{6}$
3. $\sqrt[3]{5}-8\sqrt[3]{5}$
4. $-2\sqrt{10}-7\sqrt{10}$

Answers

1. $8\sqrt{3}$
2. $\frac{2}{3}\sqrt{6}$
3. $-5\sqrt[3]{5}$
4. $-9\sqrt{10}$

Multiplication and Division

If a radical is written immediately after another radical, multiplication is intended. Sometimes a dot is placed between the radicals, but not always. Thus, either $\sqrt{7} \times \sqrt{11}$ or $\sqrt{7}\sqrt{11}$ means multiplication.

When multiplication or division of radicals is indicated, several radicals having the same index can be combined into one radical, if desired. Radicals having the same index are said to be of the Same Order. For example, a is a radical of the second order. The radicals $\sqrt{2}$ and $\sqrt{5}$ are of the same order.

If radicals are of the same order, the radicands can be multiplied or divided and placed under one radical symbol. For example, $\sqrt{5}$ multiplied by $\sqrt{3}$ is the same as $\sqrt{5\times 3}$. Also, $\sqrt{6}$ divided by $\sqrt{3}$ is the same as $\sqrt{6+3}$. If coefficients appear before the radicals, they also must be included in the multiplication or division. This is illustrated in the following examples:

1. $2\sqrt{2}.3\sqrt{5}=2.\sqrt{2}.3.\sqrt{5}$

   $=2.3\sqrt{2}.\sqrt{5}=2.3\sqrt{235}=6\sqrt{10}$

2. $\frac{15\sqrt{6}}{3\sqrt{3}}=\frac{15}{3}\times\sqrt{\frac{6}{3}}=5\times\sqrt{2}=5\sqrt{2}$.

It is important to note that what we have said about multiplication and division does not apply to addition.

A typical error is to treat the expression $\sqrt{9+4}$ as if it were equivalent to $\sqrt{9}+\sqrt{4}$. These expressions cannot be equivalent, since 3 + 2 is not equivalent to $\sqrt{13}$.

FACTORING RADICALS

A radical can be split into two or more radicals of the same order if the radicand can be factored. This is illustrated in the following examples:

1. $\sqrt{20}=\sqrt{4}.\sqrt{5}=2\sqrt{5}$
2. $\sqrt[3]{54}=\sqrt[3]{27.2}=\sqrt[3]{27}\sqrt[3]{2}=3\sqrt[3]{2}$
3. $\frac{\sqrt{20}}{\sqrt{5}}=\frac{\sqrt{4}.\sqrt{5}}{\sqrt{5}}=\sqrt{4}=2$

SIMPLIFYING RADICALS

Some radicals may be changed to an equivalent form that is easier to use. A radical is in its simplest form when no factor can be removed from the radical, when there is no fraction under the radical sign, and when the index of the root cannot be reduced.

A factor can be removed from the radical if it occurs a number of times equal to the index of the root. The following examples illustrate this:

1. $\sqrt{28}=\sqrt{2^2.7}=2\sqrt{7}$
2. $\sqrt[3]{54}=\sqrt[3]{3^2.2}=3\left(\sqrt[3]{2}\right)$
3. $\sqrt[5]{160}=\sqrt[5]{2^5.5}=3\left(\sqrt[5]{5}\right)$

Removing a factor that occurs a number of times equal to the index of the root is equivalent to separating a radical into two radicals so that one radicand is a perfect power.

The radical sign can be removed from the number that is a perfect square, cube, fourth power, etc. The root taken becomes the coefficient of the remaining radical.

In order to simplify radicals easily, it is convenient to know the squares of whole numbers up to about 25 and a few of the smaller powers of the numbers 2, 3, 4, 5, and 6.

Table. Shows Some Frequently used Powers of Numbers Table: Powers of numbers

$1^2 = 1$	$14^2 = 196$
$2^2 = 4$	$15^2 = 25$
$3^2 = 9$	$16^2 = 256$
$4^2 = 16$	$17^2 = 289$
$5^2 = 25$	$18^2 = 324$
$6^2 = 36$	$19^2 = 361$
$7^2 = 49$	$20^2 = 400$
$8^2 = 64$	$21^2 = 441$
$9^2 = 81$	$22^2 = 484$
$10^2 = 100$	$23^2 = 529$
$11^2 = 121$	$24^2 = 579$
$12^2 = 144$	$25^2 = 625$
$13^2 = 169$	

(B)	(C)
$2^1 = 2$	
$2^2 = 4$	
$2^3 = 8$	$3^1 = 3$
$2^4 = 16$	$3^2 = 9$
$2^5 = 32$	$3^3 = 27$
$2^6 = 64$	$3^4 = 81$
$2^7 = 128$	$3^5 = 243$
$2^8 = 256$	

$4^1 = 4$	$5^1 = 5$
$4^2 = 16$	$5^2 = 25$
$4^3 = 64$	$5^3 = 125$
$4^4 = 256$	$5^4 = 625$
(*D*)	(*E*)

$6^1 = 6$
$6^2 = 36$
$6^3 = 216$
$5^4 = 625$
(*F*)

Referring to table, we see that the series of numbers

1, 4, 9, 16, 25, 36, 49, 64, 81, 100

comprises all the perfect squares from 1 to 100 inclusive. If any one of these numbers appears under a square root symbol, the radical sign can be removed immediately. This is illustrated as follows:

$$\sqrt{25} = 5$$

$$\sqrt{81} = 9$$

A radicand such as 75, which has a perfect square (25) as a factor, can be simplified as follows:

$$\sqrt{75} = \sqrt{25.3} = \sqrt{25}.\sqrt{3} = 5\sqrt{3}$$

This procedure is further illustrated in the following problems:

1. $\sqrt{8} = \sqrt{4.2} = \sqrt{4}.\sqrt{2} = 2\sqrt{2}$
2. $\sqrt{72} = \sqrt{36.2} = \sqrt{36}.\sqrt{2} = 6\sqrt{2}$

By reference to the perfect fourth powers in table 7- 1, we may simplify a radical such as. Noting that $\sqrt{405}$ has the perfect fourth power 81 as a factor, we have the following:

$$\sqrt[4]{405} = \sqrt[4]{81.5} = \sqrt[4]{81}.\sqrt[4]{5} = 3\left(\sqrt[4]{5}\right)$$

As was shown with fractional exponents, taking a root is equivalent to dividing the exponent of a power by the index of the root. If a factor of the radicand has an exponent that is not a multiple of the index of the root, the factor may be separated so that one exponent is divisible by the index, as in

$$\sqrt{3^7} = \sqrt{3^6.3} = 3^{6/2}.3^{1/2} = 3^3.\sqrt{3} = 27\sqrt{3}$$

Consider also

$$\sqrt{2^3.3^7.5} = \sqrt{2^2.2.3^6.3.5} = 2.3^3\left(\sqrt{2.3.5}\right) = 54\sqrt{30}$$

If the radicand is a large number, the perfect powers that are factors are not always obvious. In such a case the radicand can be separated into prime factors. For example,

$$\sqrt{8,820} = \sqrt{2^2.3^2.5.7^2} = 2.3.7\sqrt{5} = 42\sqrt{5}$$

Practice problems: Simplify the radicals and reduce to lowest terms:

1. $\dfrac{\sqrt{3}.\sqrt{15}}{\sqrt{5}}$
2. $\dfrac{\sqrt[3]{81}}{\sqrt[3]{27}}$
3. $\dfrac{18\left(\sqrt[3]{30}\right)}{3\left(\sqrt[3]{10}\right)}$
4. $\dfrac{\sqrt{8,820}}{\sqrt{180}}$

Answers:

1. 3
2. $\sqrt[3]{3}$
3. $3\left(\sqrt[3]{3}\right)$
4. 7

RATIONAL AND IRRATIONAL NUMBERS

Real and imaginary numbers make up the number system

of algebra. Imaginary numbers are discussed in chapter 15 of this course. Real numbers are either rational or irrational. The word Rational comes from the word "ratio." A number is rational if it can be expressed as the quotient, or ratio, of two whole numbers. Rational numbers include fractions like 2/7, whole numbers, and radicals if the radical sign is removable.

Any whole number is rational. Its denominator is 1. For instance, 8 equals $\frac{8}{1}$, which is the quotient of two integers. A number like $\sqrt{16}$ is rational, since it can be expressed as the quotient of two integers in the form $\frac{4}{1}$. The following are also examples of rational numbers:

1. $\sqrt{\frac{25}{9}}$, which equals $\frac{5}{3}$
2. –6, which equals $\frac{-6}{1}$
3. $5\frac{2}{7}$, which equals $\frac{37}{7}$

Any rational number can be expressed as the quotient of two integers in many ways. For example,

$$7 = \frac{7}{1} = \frac{14}{2} = \frac{21}{3} \ldots$$

An Irrational number is a real number that cannot be expressed as the ratio of two integers. The numbers $\sqrt{3}, 5\sqrt{2}, \sqrt{7}, \frac{3}{8}\sqrt{20}$, and $\frac{2}{\sqrt{5}}$ are examples of irrational numbers.

Rationalizing Denominators

Expressions such as $\sqrt{2}$ and $5\sqrt{3}$ have irrational numbers in the denominator. If the denominators are changed immediately to decimals, as in

$$\frac{7}{\sqrt{2}} = \frac{7}{1.4142}$$

the process of evaluating a fraction becomes an exercise in long division. Such a fraction can be evaluated quickly by first changing the denominator to a rational number. Converting a fraction with an irrational number in its denominator to an equivalent fraction with a rational number in the denominator is called Rationalizing the Denominator.

Multiplying a fraction by 1 leaves the value of the fraction unchanged. Since any number divided by itself equals 1, it follows, for example, that

If the numerator and denominator of $\frac{7}{\sqrt{2}}$ are each multiplied by $\sqrt{2}$, another fraction having the same value is obtained. The result is

$$\frac{7}{\sqrt{2}} = \frac{7}{\sqrt{2}}\frac{\sqrt{2}}{\sqrt{2}} = \frac{7\sqrt{2}}{2}$$

The denominator of the new equivalent fraction is 2, which is rational. The decimal value of the fraction is

$$\frac{7\sqrt{2}}{2} = \frac{7(1.4142)}{2} = 7(0.7071) = 4.9497$$

To rationalize the denominator in $\frac{\sqrt{2}}{5\sqrt{3}}$ we multiply the numerator and denominator by $\sqrt{3}$. We get

$$\frac{\sqrt{2}}{5\sqrt{3}} = \frac{\sqrt{2}}{5\sqrt{3}}.\frac{\sqrt{3}}{\sqrt{3}} = \frac{\sqrt{6}}{5(3)} = \frac{\sqrt{6}}{15} \text{ or } \frac{1}{15}\sqrt{6}$$

Practice problems: Rationalize the denominator in each of the following:

1. $\frac{6}{\sqrt{2}}$

2. $\frac{\sqrt{5}}{\sqrt{3}}$

3. $\frac{2}{\sqrt{6}}$

4. $\frac{6}{\sqrt{y}}$

Answers:

1. $3\sqrt{2}$

2. $\frac{\sqrt{15}}{3}$

3. $\frac{\sqrt{6}}{3}$

4. $\frac{6\sqrt{y}}{y}$

EVALUATING RADICALS

Any radical expression has a decimal equivalent which may be exact if the radicand is a rational number. If the radicand is not rational, the root may be expressed as a decimal approximation, but it can never be exact. A procedure similar to long division may be used for calculating square root and cube root, and higher roots may be calculated by means of methods based on logarithms and higher mathematics. Tables of powers and roots have been calculated for use in those scientific fields in which it is frequently necessary to work with roots.

SQUARE ROOT PROCESS

The arithmetic process for calculation of square root is outlined in the following paragraphs:

1. Begin at the decimal point and mark the number off into groups of two digits each, moving both to the right and to the left from the decimal point. This may leave an odd digit at the right-hand or left-hand end of the number, or both. For example, suppose that

the number whose square root we seek is 9025. The number marked off as specified would be as follows:

$$\sqrt{9025}\,\cdot$$

2. Find the greatest number whose square is contained in the left-hand group (90). This number is 9, since the square of 9 is 81. Write 9 above the first group. Square this number (9) place its square below the left-hand group, and subtract, as follows:

$$\begin{array}{l} \quad 9 \\ \sqrt{90'25.} \\ \quad 81 \\ \hline \quad\; 9\,25 \end{array}$$

Bring down the next group (25) and place it beside the 9, as shown. This is the new dividend (925).

3. Multiply the first digit in the root (9) by 20, obtaining 180 as a trial divisor. This trial divisor is contained in the new dividend (925) five times; thus the second digit of the root appears to be 5. However, this number must be added to the trial divisor to obtain a "true divisor." If the true divisor is then too large to use with the second quotient digit, this digit must be reduced by 1. The procedure for step 3 is illustrated as follows:

$$\begin{array}{ll} & \quad 9\;\;5. \\ & \sqrt{90'25.} \\ & \quad 81 \\ 180 & \overline{\quad\; 9\,25} \\ 185 & \quad\; 9\,25 \\ & \quad\; \overline{0\,00} \end{array}$$

The number 180, resulting from the multiplication of 9 by 20, is written as a trial divisor beside the new dividend (925), as shown. The quotient digit (5) is then recorded and the trial divisor is adjusted, becoming 185. The trial quotient (180) is crossed out.

4. The true divisor (185) is multiplied by the second digit (5) and the product is placed below the new dividend (925). This step is shown in the illustration for step 3. When the product in step 4 is subtracted from the new dividend, the difference is 0; thus, in this example, the root is exact.
5. In some problems, the difference is not 0 after all of the digits of the original number have been used to form new dividends. Such problems may be carried further by adding 0's on the right-hand end of the original number, just as in normal long division. However, in the square root process the 0's must be added and used in groups of 2.

Practice problems: Find the square root of each of the following numbers:

1. 9.61
2. 123.21
3. 0.0025

Answers:

1. 3.1
2. 11.1
3. 0.05

TABLES OF ROOTS

The decimal values of square roots and cube roots of numbers with as many as 3 or 4 digits can be found from tables.

The table in appendix I of this course gives the square roots and cube roots of numbers from 1 to 100. Most of the values given in such tables are approximate numbers which have been rounded off. For example the fourth column in appendix I shows that $\sqrt{72} = 8.4853$, to 4 decimal places.

By shifting the decimal point we can obtain other square roots. A shift of two places in the decimal point in the radicand corresponds to a shift of one place in the same direction in the square root.

The following examples show the effect, as reflected in the square root, of shifting the location of the decimal point in the number whose square root we seek:

$$\sqrt{72} = 8.4853$$

$$\sqrt{0.72} = 0.84853$$

$$\sqrt{0.0072} = 0.084853$$

$$\sqrt{7,200} = 84.853$$

Cube Root

The fifth column in appendix I shows that the cube root of 72 is 4.1602. By shifting the decimal point we immediately have the cube roots of certain other numbers involving the same digits. A shift of three places in the decimal point in the radicand corresponds to a shift of one place in the same direction in the cube root.

Compare the following examples:

$$\sqrt[3]{72} = 4.1602$$

$$\sqrt[3]{0.072} = 0.41602$$

$$\sqrt[3]{72,000} = 41.602$$

Many irrational numbers in their simplified forms involve and Since these radicals occur often, it is convenient to remember their decimal equivalents as follows:

$$\sqrt{2} = 1.4142 \text{ and } \sqrt{3} = 1.7321$$

Thus any irrational numbers that do not contain any radicals other than or can be converted to decimal forms quickly without referring to tables.

For example consider

Keep in mind that the decimal equivalents of and as used in the foregoing examples are not exact numbers and the results obtained with them are approximate in the fourth decimal place.

Chapter 4

Logarithms

Logarithms 'represent a specialized use of exponents. By means of logarithms, computation with large masses of data can be greatly simplified. For example, when logarithms are used, the process of multiplication is replaced by simple addition and division is replaced by subtraction. Raising to a power by means of logarithms is done in a single multiplication, and extracting a root reduces to simple division.

DEFINITIONS

In the expression $2^3 = 8$, the number 2 is the base (not to be confused with the base of the number system), and 3 is the exponent which must be used with the base to produce the number 8. The exponent 3 is the logarithm of 8 when the base is 2. This relationship is usually stated as follows: The logarithm of 8 to the base 2 is 3. In general, the logarithm of a number N with respect to a given base is the exponent which must be used with the base to produce N. Table 8-l illustrates this.

Table: Logarithms with various bases.

$2^3 = 8$	$\log_2 8 = 3$
$4^2 = 16$	$\log_4 16 = 2$
$5^0 = 1$	$\log_5 1 = 0$
$27^{2/3} = 9$	$\log_{27} 9 = 2/3$

Table shows that the logarithmic relationship may be expressed equally well in either of two forma; these are the exponential form and the logarithmic form. Observe, that the base of a logarithmic expression is indicated by placing a subscript just below and to the right of the abbreviation 'log." Observe also that the word 'logarithm" is abbreviated without using a period. The equivalency of the logarithmic and exponential forms may be used to restate the fundamental definition of logarithms in its most useful form, as follows:

$$b^X = N \text{ implies that } \log_b N = x$$

In words, this definition is stated as follows: If the base b raised to the x power equals N, then x is the logarithm of the number N to the base b. One of the many uses of logarithms may be shown by an example in which the base is 2. Table 8-2 shows the powers of 2 from 0 through 20. Suppose that we wish to use logarithms to multiply the numbers 512 and 256, as follows:

From table 8-2, $512 = 2^9$

$$256 = 2^8$$

Then $512 \times 256 = 2^9 \times 2^8 = 2^{17}$ and from the table again $2^{17} = 131072$. It is seen that the problem of multiplication is reduced to the simple addition of the exponents 9 and 8 and finding the corresponding power in the table. Table (A) shows the base 2 in the exponential form with its corresponding powers. The actual computation in logarithmic work does not require that we record the exponential form. All that is required is that we add the appropriate exponents and have available a table in which we can look up the number corresponding to the new exponent after adding. Therefore, table (B) is adequate for our purpose. Solving the foregoing example by this table, we have the following:

$$\log_2 512 = 9$$
$$\log_2 256 = 8$$

$\log_2$ of the product = 17

Therefore, the number we seek is the one in the table whose logarithm is 17. This number is 131,072. In this example, we found the exponents directly, added them since this was a Multiplication problem, and located the correspending power. This avoided the unnecessary step of writing the base 2 each time.

Practice problems: Use the logarithms in table to perform the following multiplication:

1. 64×128

2. $1{,}024 \times 256$
3. $128 \times 4{,}096$
4. $512 \times 2{,}048$

Answers:

1. 8,192
2. 262,144
3. 524,288
4. 1,048,576

NATURAL AND COMMON LOGARITHMS

Many natural phenomena, such as rates of growth and decay, are most easily described in terms of logarithmic or exponential formulas.

Furthermore, the geometric patterns in which certain seeds grow (for example, sunflower seeds) is a logarithmic spiral.

These facts explain the name "natural logarithms." Natural logarithms use the base e, which is an irrational number approximately equal to 2.71828. This system is sometimes called the Napierian system of logarithms, in honor of John Napier, who is credited with the invention of logarithms.

To distinguish natural logarithms from other logarithmic systems the abbreviation, In, is sometimes used. When In appears, the base is understood to be e and need not be shown. For example, either log, 45 or In 45 signifies the natural logarithm of 45.

COMMON LOGARITHMS

As has been shown in preceding paragraphs, any number may be used as a base for a system of logarithms.

The selection of a base is a matter of convenience. Briggs in 1617 found that base 10 possessed many advantages not obtainable in ordinary calculations with other bases. The

Table: Exponential and logarithmic tables for the base 2

(A) Powersof 2 from
(B) Logarithms for the base 2 and
0through 20 corres ponding powers

	Log	Number
$2^0 = 1$	0	1
$2^1 = 2$	1	2
$2^2 = 4$	2	4
$2^3 = 8$	3	8
$2^4 = 1$	4	16
$2^5 = 128$	6	128
$2^6 = 64$	6	64
$2^7 = 128$	7	128
$2^8 = 258$	8	256
$^{2}9 = 512$	9	512
$2^{10} = 1042$	10	1024
$2^{11} = 2048$	11	2048
$2^{12} = 4096$	12	4096
$2^{13} = 8192$	13	8192
$2^{14} = 16384$	14	16384
$2^{15} = 32768$	16	65536
$2^{16} = 65536$	17	131072
$2^{17} = 131072$	18	262144
$2^{18} = 262144$	18	262144
$2^{19} = 524288$	19	524288
$2^{20} = 1048578$	20	1048576

selection of 10 as a base proved so satisfactory that today it is used almost exclusively for ordinary calculations.

Logarithms with 10 as a base are therefore called Common Logarithms. When 10 is used as a base, it is not necessary to indicate it in writing logarithms. For example,

$$\log 100 = 2$$

is understood to mean the same as

$$\log_{10} 100 = 2$$

If the base is other than 10, it must be specified by the use of a subscript to the right and below the abbreviation "log."

As noted in the foregoing discussion of natural logarithms, the use of the distinctive abbreviation "In" eliminates the need for a subscript when the base is *e*.

It is relatively easy to convert common logarithms to natural logarithms or vice versa, if necessary.

It should be noted further that each system has its peculiar advantages, but for most everyday work, the common system is more often used. A simple relation connects the two systems. If the common logarithm of a number can be found, multiplying by 2.3026 gives the natural logarithm of the number. For example,

$$\log 1.60 = 0.2041$$

$$\ln 1.60 = 2.3026 \times 0.2041 = 0.4700$$

Thus the natural logarithm of 1.60 is 0.4700, correct to four significant digits.

Conversely, multiplying the natural logarithm by 0.4343 gives the common logarithm of a number. As might be expected, the conversion factor 0.4343 is the reciprocal of 2.3026. This is shown as follows:

$$\frac{1}{2.30626} = 0.4343$$

Positive Integral Logarithms

The derivation of positive whole logarithms is readily apparent. For example, we see in table (B) that the logarithm of 10 is 1.

The number 1 is simply the exponent of the base 10 which yields 10. This is shown in table (A) opposite the logarithmic equation. Similarly,

$10^0 = 1$ .... $\log 1 = 0$

$10^2 = 100$ .... $\log 100 = 2$

$10^3 = 1{,}000$ ...... $\log 1{,}000 = 3$

$$10^4 = 10{,}000 \text{ ..... } \log 10{,}000 = 4$$

Table. Exponential and Corresponding Logarithmic Notations Using Base 10

A	B
$10^{-4} = \frac{1}{10^4} = 0.0001$	log 0.0001 = -4
$10^{-3} = \frac{1}{10^3} = 0.001$	log 0.001 = – 3
$10^{-2} = \frac{1}{10^2} = 0.01$	log 0.01 = – 2
$10^{-1} = \frac{1}{10^2} = 0.01$	log 0.01 = – 2
$10^{-1/2} = \frac{1}{\sqrt{10}} = \frac{\sqrt{10}}{10} = 0.31623$	log 0.31623
	= – 0.5 = 0.5 – 1
$10^9 = 1$	log 1 = 0
$10^{1/2} = \sqrt{10} = 3.1623$	log 3.1623 = 0.5
$10^1 = 10$	log 10 = 1
$10^{3/2} = 10\sqrt{10} = 31.623$	log 31.623 = 1.5
$10^2 = 100$	log 100 = 2
$10^{5/2} = 10^2\left(\sqrt{10}\right) = 316.23$	log 316.23 = 2.5
$10^3 = 1{,}000$	log 1,000 = 3
$10^{7/2} = 10^3\left(\sqrt{10}\right) = 3162.3$	log 3162.3 = 3.5
$10^4 = 10{,}000$	log 10,000 = 4

Positive Fractional Logarithms

Referring to table 8-3, notice that the logarithm of 1 is 0 and the logarithm of 10 is 1. Therefore, the logarithm of a number between 1 and 10 is between 0 and 1. An easy way to verify this is to consider some numbers between 1 and 10 which are powers of 10; the exponent in each case will then be the logarithm we seek. Of course, the only powers of 10 which produce numbers between 1 and 10 are fractional powers.

Example: $10^{1/2} = 3.1623$ (approximately)

$$10^{0.5} = 3.1623$$

Therefore, log 3.1623 = 0.5

Other examples are shown in the table for $10^{3/2}$, $10^{5/2}$, and $10^{7/2}$. Notice that the number that represents $10^{3/2}$, 31.623, logically enough lies between the numbers representing 10^1 and 10^2 -that is, between 10 and 100. Notice also that $10^{5/2}$ appears between 10^2 and 10^3, and $10^{7/2}$ lies between lo3 and 104.

Negative Logarithms

Table shows that negative powers of 10 may be fitted into the system of logarithms. We recall that 10^{-1} means 1/10, or the decimal fraction, 0.1. What is the logarithm of 0.1?

Solution: $10^{-1} = 0.1$; log 0.1 = –1

Likewise $10^{-2} = 0.01$; log 0.01 = –2

Negative Fractional Logarithms

Notice in table that negative fractional exponents present no new problem in logarithmic notation. For example, 1O'1'2 means

$$\frac{1}{\sqrt{10}}$$

$$\frac{1}{\sqrt{10}} = \frac{\sqrt{10}}{10} = 0.31623$$

What is the logarithm of 0.316231

Solution:

$$10^{-1/2} = 0.31623; \log 0.31623 = -1/2 = -0.5$$

Table shows logarithms for numbers ranging from 0.0001 to 10,000. Notice that there are only 8 integral logarithms in the entire range. Excluding zero logarithms, the logarithms for all other numbers in the range are fractional or contain a fractional part. By the year 1628, logarithms for all integers from 1 to 100,000 had been computed.

Practically all of these logarithms contain a fractional part. It should be remembered that finding the logarithm of a number is nothing more than expressing the number as a power of 10. Table shows the numbers 1 through 10 expressed as powers of 10. Most of the exponents which comprise logarithms are found by methods be- yond the scope of this text. However, it is not necessary to know the process used to obtain logarithms in order to make use of them.

Table: The Numbers 1 through 10 Expressed as Powers of 10

$1 = 10^0$	$6 = 10^{0.77815}$
$2 = 10^{0.30103}$	$7 = 10^{0.84510}$
$3 = 10^{0.47712}$	8 = 100.90309
$4 = 10^{0.60206}$	$9 = 10^{0.95424}$
$5 = 10^{0.69697}$	$10 = 10^1$

COMPONENTS OF LOGARITHMS

The fractional part of a logarithm is usually written as a decimal. The whole number part of a logarithm and the decimal part have been given separate names because each plays a special part in relation to the number which the logarithm represents. The whole number part of a logarithm is called the Characteristic. This part of the logarithm shows the position of the decimal point in the associated number. The decimal part of a logarithm is called the Mantissa.

For a particular sequence of digits making up a number, the mantissa of a common logarithm is always the same regardless of the position of the decimal point in that number. For example, log 5270 = 3.72181; the mantissa is 0.72181 and the characteristic is 3.

CHARACTERISTIC

The characteristic of a common logarithm shows the position of the decimal point in the associated number. The characteristic for a given number may be determined by inspection. It will be remembered that a common logarithm is simply an exponent of the base 10. It is the power of 10 when a number is written in scientific notation.

When we write log 360 = 2.55630, we understand this to mean 10 2.55630 = 360. We know that the number is 360 and not 36 or 3,600 be cause the characteristic is 2. We know 10 is 10, 10^2 is 100, and 10^2 is 1,000. Therefore, the number whose value is $10^{2.55630}$ must lie between 100 and 1,000 and of course any number in that range has 3 digits.

Suppose the characteristic had been 1: where would the decimal point in the number be placed? Since 10^1 is 10 and 10^2 is 100, any number whose logarithm is between 1 and 2 must lie between 10 and 100 and will have 2 digits. Notice how the position of the decimal point changes with the value of the characteristic in the following examples:

$$\log 36{,}000 = 4.55630$$
$$\log 3{,}600 = 3.55630$$
$$\log 360 = 2.55630$$
$$\log 36 = 1.55630$$
$$\log 3.6 = 0.55630$$

Note that it is only the characteristic that changes when the decimal point is moved. An advantage of using the base 10 is thus revealed: If the characteristic is known, the decimal point may easily be placed. If the number is known, the characteristic may be determined by inspection; that is, by observing the location of the decimal point.

Although an understanding of the relation of the characteristic to the powers of 10 is necessary for thorough comprehension of logarithms, the characteristic may be determined mechanically by application of the following rules:

1. For a number greater than 1, the characteristic is positive and is one less than the number of digits to the left of the decimal point in the number.
2. For a positive number less than 1, the characteristic is negative and has an absolute value one more than

the number of zeros between the decimal point and the first nonzero digit of the number.

Table contains examples of each type of characteristic.

Table: Positive and Negative Characteristics

Number	Power of 10	Digits in number	Characteristic to the left of decimal point
		Between:	
134	10^2 and 10^3	3	2
13.4	10^1 and 10^2	2	1
1.34	10^0 and 10^1	1	0
			Zeros between decimal point and first non zero digit
0.134	10^{-1} and 10^0	0	–1
0.0134	10^{-2} and 10^{-1}	1	–2
0.00134	10^{-3} and 10^{-2}	2	–3

Practice problems: In problems 1 through 4, write the characteristic of the logarithm for each number.

In 5 through 8, place the decimal point in each number as indicated by the characteristic (c) given for each.

1. 4,321
2. 1.23
3. 0.05
4. 12
5. 123; c = 4
6. 8,210; c = 0
7. 8; c = –1
8. 321; c = –2

Answers:

1. 3

2. 0
3. –2
4. 1
5. 12,300
6. 8.210
7. 0.8
8. 0.0321

Negative Characteristics

When a characteristic is negative, such as –2, we do not carry out the subtraction, since this would involve a negative mantissa.

There are several ways of indicating a negative characteristic. Mantissas as presented in the table in the appendix are always positive and the sign of the characteristic is indicated separately.

For example, where log 0.023 = 2.36173, the bar over the 2 indicates that only the characteristic is negative-that is, the logarithm is –2 + 0.36173.

Another way to show the negative characteristic is to place it after the mantissa. In this case we write 0.36173–2.

A third method, which is used where possible throughout this chapter, is to add a certain quantity to the characteristic and to subtract the same quantity to the right of the mantissa. In the case of the example, we may write:

$$\begin{array}{l} \bar{2}.36173 \\ \underline{10 \qquad -10} \\ 8.36173 - 10 \end{array}$$

In this way the value of the logarithm remains the same but we now have a positive characteristic as well as a positive mantissa.

MANTISSA

The mantissa is the decimal part of a logarithm. Tables of

logarithms usually contain only mantissas since the characteristic can be readily determined as explained previously.

Table 8-8 shows the characteristic, mantissa, and logarithm for several positions of the decimal point using the sequence of digits 4, 5, 6.

It will be noted that the mantissa remains the same for that particular sequence of digits, regardless of the position of the decimal point.

Table: Effect of Changes in the Location of the Decimal Point

Number	*Characteristics*	*Mantissa*	*Logarithm*
45,600	4	0.6590	4.6590
4.560	3	0.6590	3.6590
456	2	0.6590	2.6590
45.6	1	0.6590	1.6590
4.56	–1	0.6590	0.6590 – 1
0.0456	–2	0.6590	0.6590 – 2
0.00456	–3	0.6590	0.6590 – 3

Appendix I of this training course is a table which includes the logarithms of numbers from 1 to 100. For our present purpose in using this table, we are concerned only with the first and sixth columns.

The first column contains the number and the sixth column contains its logarithm. For example, if it is desired to find the logarithm of 45, we would find the number 45 in the first column, look horizontally across the page to column 6 and read the logarithm, 1.65321. A glance down the logarithm column will reveal that the logarithms increase in value as the numbers increase in value. It must be noted in this particular table that both the mantissa and the characteristic are given for the number in the first column.

This is simply an additional aid, since the characteristic can easily be determined by inspection. Suppose that we wish to use the table of Appendix I to find the logarithm of a number not shown in the "number" column. By recalling that the mantissa does not change when the decimal point moves, we may be able to determine the desired logarithm. For example, the number 450 does not appear in the number column of the table. However, the number 45 has the same mantissa as 450; the only difference between the two logs is in their characteristics. Thus the logarithm of 450 is 2.65321.

Practice problems: Find the logarithms of the following numbers:

1. 64
2. 98
3. 6400
4. 9.8

Answers:

1. 1.80618
2. 1.99123
3. 3.80618
4. 0.99123

THE SLIDE RULE

In 1620, not long after the invention of logarithms, Edmond Gunter showed how logarithmic calculations could be carried out mechanically. This is done by laying off lengths on a rule, representing the logarithms of numbers, and by combining these lengths in various ways. The idea was developed and with the contributions of Mannheim in 1851 the slide rule came into being as we know it today.

The slide rule is a mechanical device by which we can carry out any arithmetic calculation with the exception of addition and subtraction. The most common operations with the slide rule are multiplication, division, finding the square or cube of a number, and finding the square root or cube root of a number. Also trigonometric operations are frequently performed. The advantage of the slide rule is that it can be used with relative ease to solve complicated problems. One limitation is that it will give results with a maximum of only three accurate significant digits. This is sufficient in most calculations, however, since most physical constants are only correct to two or three significant digits. When greater accuracy is required, other methods must be used. A simplified diagram of a slide rule is pictured in figure.

The sliding, central part of the rule is called the Slide. The movable glass or plastic runner with a hairline imprinted on it is called the Indicator. There is a C scale printed on the slide, and a D scale exactly the same as the C scale printed on the Body or Stock of the slide rule. The mark that is associated with the primary number 1 on any slide rule scale is called the Index. There is

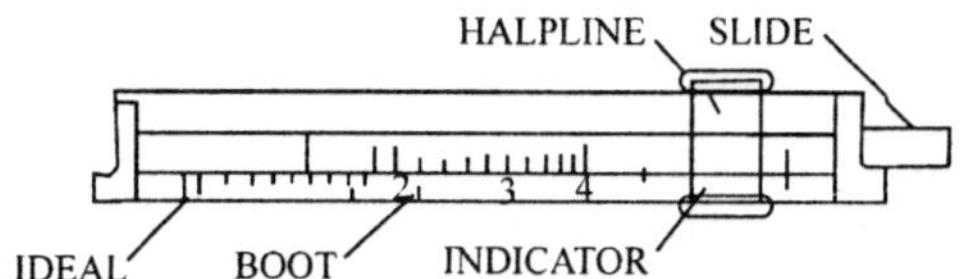

Fig. Simplified Diagram of a slide rule an Index at the Extreme left and at the Extreme right on both the C and D scales

SLIDE RULE THEORY

We have mentioned that the slide rule is based on logarithms. Recall that, to multiply two numbers, we simply add their logarithms. Previously we found these iogarithms in tables, but if the logarithms are laid off on scales such as the C and D scale of the slide rule, we can add the lengths, which represent these logarithms. To make such a scale we could mark off mantissas ranging from 0 to 1 on a rule as in figure. We then find in the tables the logarithms for numbers ranging from 1 to 10 and write the number opposite its corresponding logarithm on the scale.

MIRROR 1 2 3 4 5 6 7 8 9 0

LOCATION 0 0.1 0.2 0.3 0.4 0.5 0.6 0.7 0.8 0.9 1.0

Fig. Logarithms and Corresponding Numbers on a Scale

Table lists the numbers 1 through 10 and their corresponding logarithms to three places. These numbers are written opposite their logarithms on the scale shown in figure 8-2. If we have two such scales, exactly alike, arranged so that one of them is free to slide along the other, we can perform the operation of multiplication, for example, by Adding Lengths that is, by adding logarithms.

For example, if we wish to multiply 2 x 3, we find the logarithm of 2 on the stationary scale and move the sliding scale so that its index is over that mark.

We then add the logarithm of 3 by finding that logarithm on the sliding scale and by reading below it, on the stationary scale, the logarithm that is the sum of the two.

Since we are not interested in the logarithms themselves, but rather in the numbers they represent, it is possible to remove the logarithmic notation on the scale in figure, and leave only the logarithmically spaced number scale.

The C and D scales of the ordinary slide rule are made up in this manner. Figure shows the multiplication of 2 x 3. Although the logarithm scales have been removed, the numbers 2 and 3 in reality signify the logarithms of

Table: Numbers and Their Corresponding Logarithms

Number	*Logarithm*	*Number*	*Logarithm*
1	0.000	6	0.778
2	0.301	7	0.845
3	0.477	8	0.903
4	0.602	9	0.945
5	0.699	10	1.000

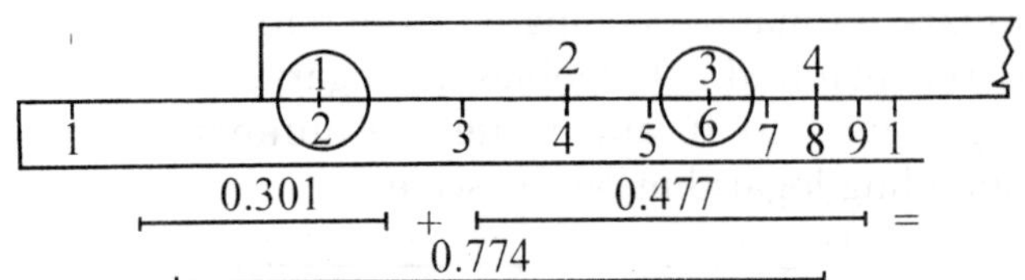

Fig. Multiplication by use of the slide rule

2 and 3, namely, 0.301 and 0.477; the product 6 on the scale really signifies the logarithm of 6, that is, 0.778. Thus, although logarithms are the underlying principle, we are able to work with the numbers directly.

It should be noted that the scale is made up from mantissas only. The characteristic must be determined separately as in the case where tables are used. Since mantissas identify only the digit sequence, the digit 3 on the slide rule represents not only 3 but 30, 300, 0.003, 0.3, and so forth.

Thus, the divisions may represent the number multiplied or divided by any power of 10. This is true also for numbers that fall between the divisions. The digit sequence, 1001, could represent 100.1, 1.001, 0.01001, and so forth. The following example shows the use of the same set of mantissas which appear in the foregoing example, but with a different characteristic and, therefore, a different answer:

Example: Use logs (positions on the slide rule) to multiply 20 times 30.

Solution:

log 20 = 1.301 (2 on the slide rule)

log 30 = 1.477 (3 on the slide rule)

log of answer = 2.778 (6 on the slide rule)

Since the 2 in the log of the answer is merely the indicator of the position of the decimal point in the answer itself, we do not expect to find it on the slide rule scale.

As in the foregoing example, we find the digit 6 opposite the multiplier 3. This time, however, the 6 represents 600, because the characteristic of the log represented by 6 in this problem is 2.

READING THE SCALES

Reading a slide rule is no more complicated than reading a yard stick or ruler, if the differences in its markings are understood.

Between the two indices of the C or D scales (the large digit 1 at the extreme left and right of the scales) are divisions numbered 2, 3, 4,5, 6, 7,8, and 9. Each length between two consecutive divisions is divided into 10 sections and each section is divided into spaces.

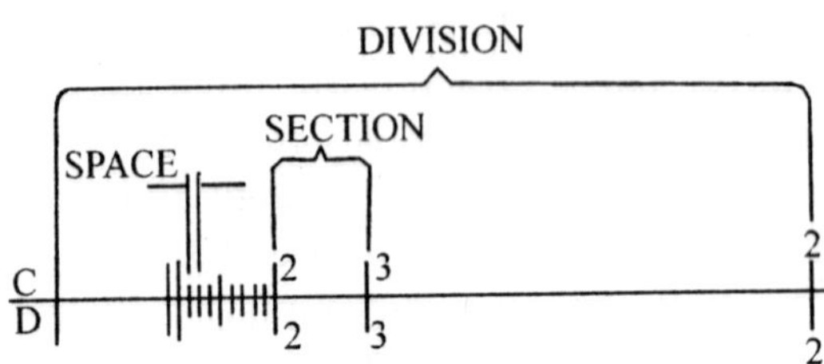

Fig. Division, Section, and Space of a Slide Rule Scale

Notice that the division between 1 and 2 occupies about one-third of the length of the rule. This is sufficient space in which to write a number for each of the section marks. The sections in the remaining divisions are not numbered, because the space is more limited. Notice also that in the division between 1 and 2, the sections are each divided into 10 spaces.

The sections of the divisions from 2 to 4 are subdivided into only 5 spaces, and those from 4 to the right index are subdivided into only 2 spaces. These subdivisions are so arranged because of the limits of space.

Only the sequence of significant digits is read on the slide rule. The position of the decimal point is determined separately. For example, if the hairline of the indicator is in the left-hand position shown in figure, the significant digits are read as follows:

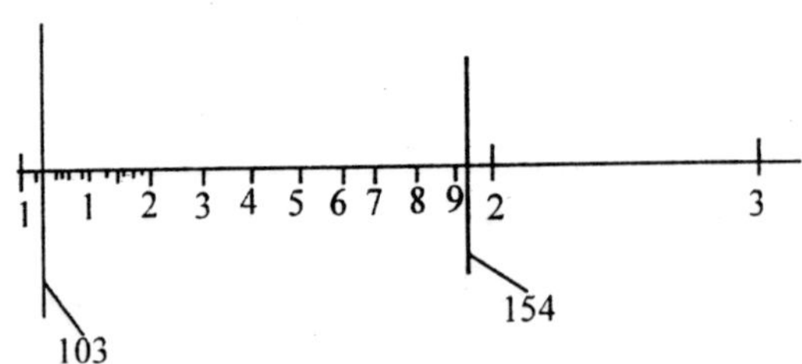

Fig. Readings in the first Division of a Slide Rule

1. Any time the hairline falls in the first division, the first significant digit is 1.
2. Since the hairline lies between the index and the first section mark, we know the number lies between 1.0

and 1.1, or 10 and 11, or 100 and 110, etc. The second significant digit is 0.

3. We next find how far from the index the hairline is located. It lies on the marking for the third space.
4. The three significant digits are 103. In the second example shown in figure 8-5, the hairline is located in the first division, the ninth section, and on the fourth space mark of that section. Therefore, the significant digits are 194. number from 0 through 9 as its second digit and any number from 0 through 9 as its third digit. Sometimes a fourth digit can be roughly approximated in this first division, but the number is really accurate to only three significant digits.

In the second and third divisions, each section is divided into only 5 spaces. Thus, each space is equal to 0.2 of the section. Suppose, for example, that the hairline lies on the third space mark after the large 2 indicating the second division.

The first significant digit is 2. Since the hairline lies between 2 and the first section mark, the second digit is 0. The hairline lies on the third space mark or 0.6 of the way between the division mark and the first section mark, so the third digit is 6. Thus, the significant digits are 206. Notice that if the hairline lies on a space mark the third digit can be written accurately; otherwise it must be approximated. Thus, we see that any number falling in the first division of the slide rule will always have 1 as its first significant digit. It can have any

From the fourth division to the right index, each section is divided into only.two spaces. Thus, if the hairline is in the fourth division and lies on the space mark between the sixth and seventh sections, we would read 465. If the hairline did not fall on a space mark, the third digit would have to be approximated.

OPERATIONS WITH THE SLIDE RULE

There are two parts in solving problems with' a slide rule.

In the first part the slide rule is used to find the digit sequence of the final result. The second part is concerned with the placing of the decimal point in the result. Let us consider first the digit sequence in multiplication and division.

Multiplication

Multiplication is performed on the C and D scales of the slide rule. The following procedure is used:

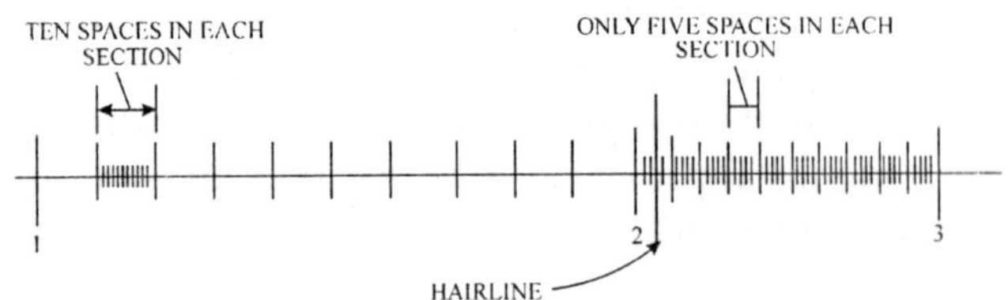

Fig. Reading in the Second Division of a Slide Rule

1. Locate one of the factors to be multiplied on the D scale, disregarding the decimal point.
2. Place the index of the C scale opposite that number.
3. Locate the other factor on the C scale and move the hairline of the indicator to cover this factor.
4. The.product is on the D scale under the hairline.

Sometimes in multiplying numbers, such as 25 x 6, the number on the C scale extends to the right of the stock and the product cannot be read. In such a case, we simply shift indices. Instead of the left-hand index of the C scale, the right-hand index is placed opposite the factor on the D scale. The rest of the problem remains the same. By shifting indices, we are simply multiplying or dividing by 10, but this plays no part in reading the significant digits. Shifting indices affects the characteristic only.

Example: 252 × 3 = 756

1. Place the left index of the C scale over 252.
2. Locate 3 on the C scale and set the hair- line of the indicator over it.

3. Under the hairline on the D scale read the product, 756.

Example: 4 × 64 = 256

1. Place the right index of the C scale over 4.
2. Locate 64 on the C scale and set the hairline of the indicator over it.
3. Under the hairline on the D scale read the product, 256.

Practice problems: Determine the following products by slide rule to three significant digits:

1. 2.8 × 16
2. 7 × 1.3
3. 6 × 85
4. 2.56 × 3.5

Answers:

1. 44.8
2. 9.10
3. 510
4. 8.96

Division

Division being the inverse of multiplication, the process of multiplication is reversed to perform division on a slide rule. We subtract the length representing the logarithm of the divisor from the length representing the logarithm of the dividend to get the logarithm of the quotient.

The procedure is as follows:

1. Locate the dividend on the D scale and place the hairline of the indicator over it.
2. Move the slide until the divisor (on the C scale) lies under the hairline.
3. Read the quotient on the D scale opposite the C scale index.

If the divisor is greater numerically than the dividend, the slide will extend to the left. If the divisor is less, the slide will extend to the right. In either case, the quotient is the number on the D scale that lies opposite the C scale index, falling within the limits of the D scale.

Example: 6 ÷ 3 = 2

1. Locate 6 on the D scale and place the hairline of the indicator over it.
2. Move the slide until 3 on the C scale is under the hairline.
3. Opposite the left C scale index, read the quotient, 2, on the D scale.

Example: 378 ÷ 63 = 6

1. Locate 378 on the D scale and move the hairline of the indicator over it.
2. Move the slide to the left until 63 on the C scale is under the hairline.
3. Opposite the right-hand index of the C scale, read the quotient, 6, on the D scale.

Practice problems: Determine the following quotients by slide rule.

1. 126 ÷ 3
2. 960 ÷ 15
3. 142 ÷ 71
4. 459 ÷ 17

Answers:

1. 42
2. 64
3. 2
4. 27

PLACING THE DECIMAL POINT

Various methods have been advanced regarding the placement of the decimal point in numbers derived from slide rule computations. Probably the most universal and most easily remembered method is that of approximation.

The method of approximation means simply the rounding off of numbers and the mechanical shifting of decimal points in the numbers of the problem so that the approximate size of the solution and the exact position of the decimal point will be seen from inspection.

The slide rule may then be used to derive the correct sequence of significant digits. The method may best be demonstrated by a few examples. Remember, shifting the decimal point in a number one place to the left is the same as dividing by 10. Shifting it one place to the right is the same as multiplying by 10. Every shift must be compensated for in order for the solution to be correct.

Example: 0.573 × 1.45

Solution: No shifting of decimals is necessary here. We see that approximately 0.6 is to be multiplied by approximately 1 1/2. Immediately, we see that the solution is in the neighborhood of 0.9. By slide rule we find that the significant digit sequence of the product is 832. From our approximation we know that the decimal point is to the immediate left of the first significant digit, 8. Thus,

$$0.573 \times 1.45 = 0.832$$

Example: 239 × 52.3

Solution: For ease in multiplying, we shift the decimal point in 52.3 one place to the left, making it 5.23. To compensate, the decimal point is shifted to the right one place in the other factor. The new position of the decimal point is indicated by the presence of the caret symbol.

$$239.0_{\wedge} \times 5_{\wedge}2.3$$

Our problem is approximately the same as

$$2{,}400 \times 5 = 12{,}000$$

By slide rule the digit sequence is 125. Thus,

$$239 \times 52.3 = 12{,}500$$

Example: 0.000134 × 0.092

Solution: Shifting decimal points, we have

$$0_{\wedge}00.000134 \times 0.09_{\wedge}2$$

Approximation: 9 × 0.0000013 = 0.0000117. By slide rule the digit sequence is 123. From approximation the decimal point is located as follows:

$$0.0000123$$

Thus,

$$0.000134 \times 0.092 = 0.0000123$$

Example: $\frac{53.1}{42.4}$

Solution: The decimal points are shifted so that the divisor becomes a number between 1 and 10. The method employed is cancellation. Shifting decimal points, we have

$$\frac{5 \wedge 3.1}{4 \wedge 2.4}$$

Approximation: 5/4 = 1.2

Digit sequence by slide rule:

$$1255$$

Placing the decimal point from the approximation:

$$1.255$$

Thus,

$$\frac{53.1}{42.4} = 1.255$$

Example:

$$\frac{0.00645}{0.0935}$$

Solution: Shifting decimal points

$$\frac{0.00 \wedge 645}{0.09 \wedge 35}$$

Approximation:

$$\frac{0.6}{9} = 0.07$$

Digit sequence by slide rule: 690

Placing the decimal point from the approximation:

$$0.0690$$

Thus,

$$\frac{0.00645}{0.0935} = 0.0695$$

Practice problems: Solve the following problems with the slide rule and use the method of approximation to determine the position of the decimal point:

Answers:

1. 0.000745
2. 42.4
3. 0.0438
4. 0.212

MULTIPLICATION AND DIVISION COMBINED

In problems such as

$$\frac{0.644 \times 330}{161 \times 12}$$

It is generally best to determine the position of the decimal point by means of the method of approximation and to determine the significant digit sequence from the slide rule. Such problems are usually solved by dividing and multiplying alternately throughout the problem. That is, we divide 0.644 by 161, multiply the quotient by 330, and divide that product by 12. Shifting decimal points, we have

$$\frac{0 \wedge 0644 \times 3 \wedge 30}{1 \wedge 61 \times 1 \wedge 2}$$

Since there is a combined shift of three places to the left in the divisor, there must also be a combined shift of three places to the left in the dividend.

Approximation: $\frac{0.06 \times \cancel{3}^{2}}{\cancel{1}.\cancel{5}}$ = 0.06 × 2 = 0.12.

The step-by-step process of determining the significant digit sequence of this problem is as follows:

1. Place the hairline over 644 on the D scale.

2. Draw the slide so that 161 of the C scale lies under the hairline opposite 644.
3. Opposite the C scale index (on the D scale) is the quotient of 644 + 161. This is to be multiplied by 330, but 330 projects beyond the rule so the C scale indices must be shifted.
4. After shifting the indices, find 330 on the C scale and place the hairline over it. Opposite 330 under the hairline on the D scale is the product of $\frac{644}{161} \times 330$.
5. Next, move the C scale until 12 is under the hairline. Opposite the C scale index (on the D scale) is the final quotient. The digit sequence is 110.

The decimal point is then placed according to our approximation: 0.11. Thus,

$$\frac{0.644 \times 330}{161 \times 12} = 0.11$$

Practice problems. Solve the following problems, using a slide rule:

1. $\frac{22 \times 78.5 \times 157}{17 \times 18.3 \times 85}$
2. $\frac{242 \times 78.5 \times 157}{25{,}600 \times 198}$
3. $\frac{2.77 \times 0.064}{0.17 \times 1.97}$

Answers:

1. 10.2
2. 0.817
3. 0.529

SQUARES

Squares of numbers are found by reference to the A scale. The numbers on the A scale are the squares of those on the D Scale.

The A scale is really a double scale, each division being one-half as large as the corresponding division on the D scale. The use of a double scale for squaring is based upon the fact that the logarithm of the square of a number is twice as large as the logarithm of the number itself. In other words,

$$\log N^2 = 2 \log N$$

This is reasonable, since

$$\log N^2 = \log (N \times N) = \log N + \log N$$

For a numerical example, suppose that we seek to square 2 by means of logarithms.

$$\log 2 = 0.301$$

$$\log 2^2 = 2 \log 2 = 2 \times 0.301 = 0.602$$

Since each part of the A scale is half as large as the corresponding part of the D scale, the logarithm 0.602 on the A scale will be the same length as the logarithm 0.301 on the D scale. That is, these logarithms will be opposite on the A and D scales. On the A scale as on the D scale, the numbers are written rather than their logarithms. Select several numbers on the D scale, such as 2, 4, 8, 11, and read their squares on the A scale, namely 4, 16, 64, 121.

Notice also that the same relation exists for the B and C scales as for the A and D scales. Of interest, also, is the fact that since the A and B scales are made up as are the C and D scales, they too could be used for multiplying or dividing.

Placing the Decimal Point

Usually the decimal may be placed by the method of approximation. However, close observation will reveal certain facts that eliminate the need for approximations in squaring numbers. Two rules suffice for squaring whole or mixed numbers, as follows:

1. When the square of a number is read on the left half of the A scale, that number will contain twice the number of digits to the left of the decimal point in the original number, less 1.
2. When the square of a number is read on the right

half of the A scale, that number will contain twice the number of digits to the left of the decimal point in the original number.

Example: Square 2.5.

Solution: Place the hairline over 25 on the D scale. Read the digit sequence, 625, under the hairline in the left half of the A scale.

By rule 1: (2xnumberof digits)–1 = 2(1)–1 = 1. There is one digit to the left of the decimal point. Thus,

$$(2.5)^2 = 6.25$$

Example: Square 6,340.

Solution: Digit sequence, right half A scale: 402.

By rule 2: 2 × number of digits = 2 × 4 = 8 (digits in answer). Thus,

$$(6{,}340)^2 = 40{,}200{,}000$$

Positive Numbers Less Than One

If positive numbers less than one are to be squared, a slightly different version of the preceding rules must be employed. Count the zeros between the decimal point and the first nonzero digit. Consider this count negative. Then the number of zeros between the decimal point and the first significant digit of the squared number may be found as follows:

1. Left half A scale: Multiply the zeros counted by 2 and subtract 1.
2. Right half A scale: Multiply the zeros counted by 2.

Example: Square 0.0045

Solution: Digit sequence, right half A scale: 2025. By rule 2: 2(–2) = –4. (Thus, 4 zeros between the decimal point and the first digit.)

$$(0.0045)^2 = 0.09002025$$

Example: Square 0.0215

Solution: Digit sequence, left half A scale: 462. By rule 1: 2(–1) –1 = –3

$$(0.0215)^2 = 0.000462$$

SQUARE ROOTS

Taking the square root of a number with the slide rule is the inverse process of squaring a number. We find the number on the A scale, set the hairline of the indicator over it, and read the square root on the D scale under the hairline.

Positioning Numbers on the A Scale Since there are two parts of the A scale exactly alike and the digit sequence could be found on either part, a question arises as to which section to use. Generally, we think of the left half of the rule as being numbered from 1 to 10 and the right half as being numbered from 10 to 100. The numbering continues- left half 100 to 1,000, right half 1,000 to 10,000, and so forth.

A simple process provides a check of the location of the number from which the root is to be taken. For whole or mixed numbers, begin at the decimal point of the number and mark off the digits to the left (including end zeros) in groups of two. This is illustrated in the following two examples:

1. $\sqrt{40,300.21}$

   $\sqrt{4'03'00.21}$

2. $\sqrt{2.034.1}$

   $\sqrt{20'34.1}$

Look at the left-hand group. If it is a l-digit number, use the left half of the A scale. If it is a a-digit number, use the right half of the A scale. The number in example 1 is thus located in the left half of the A scale and the number in example 2 is located in the right half.

Numbers Less Than One

For positive numbers less than one, begin at the decimal point and mark off groups of two to the right. This is illustrated as follows:

1. $\sqrt{0.000245}$

$\sqrt{0.00'02'45}$

2. $\sqrt{0.00402}$

$\sqrt{0.00'40'2}$

Looking from left to right, locate the first group that contains a digit other than zero. If the first figure in this group it3 zero, locate the number in the left half of the A scale. If the first figure is other than zero, locate the number in the right half of the A scale. Thus, $\sqrt{0.00'02'45}$ is located left.

and

$\sqrt{0.00'40'2}$ is located right

Powers of 10

When the square root of 10, 1,000, 100,000, and so forth, is desired, the center index is used. That is, when the number of digits in a power of 10 is even, use the center index. The slide rule uses only the first three significant digits of a number. Thus, if the rule is used, $\sqrt{23451.6}$ must be considered as $\sqrt{23500.0}$. Likewise, 1.43567 would be considered 1.43000, and so forth. For greater accuracy, other methods must be used.

Practice problems: State which half of the A scale should be used for each of the following:

1. $\sqrt{432}$
2. $\sqrt{0.014}$
3. $\sqrt{241.67}$
4. $\sqrt{0.00045}$
5. $\sqrt{4,320}$
6. $\sqrt{0.00301}$
7. $\sqrt{0,0640}$
8. $\sqrt{9.41}$

Answers:

1. Left
2. Left
3. Left
4. Left
5. Right
6. Right
7. Left
8. Left

Placing the Decimal Point

To place the decimal point in the square root of a number, mark off the original number in groups of two as explained previously.

For whole or mixed numbers, the number of groups marked off is the number of digits including end zeros to the left of the decimal point in the root. The following problems illustrate this:

1. $\sqrt{23,415}$, $\sqrt{2'34'15}$

Three digits to left of decimal point in square root

2. $\sqrt{421,562.4}\sqrt{42'15'62.4}$

Three digits to left of decimal point in square root

3. $\sqrt{231.321}\sqrt{2'31.321}$

Two digits to left of decimal point in square root

For positive numbers less than one, there will be one zero in the square root between the decimal point and the first significant digit for every pair of zeros counted between the decimal point and the first significant digit of the original number. This is illustrated as follows:

1. $\sqrt{0.0004}\sqrt{0.00'04}$

One zero before first digit in square root

2. $\sqrt{0.00008}\sqrt{0.00'00'8}$

Two zeros before first digit in square root

3. $\sqrt{0.08'}$

No zeros before first digit in square root

Example: $\sqrt{4,521}\sqrt{45'21}$

(Two digits in left-hand group)

Place the hairline over 452 on the right half of the A scale. Read the digit sequence of the root, 672, on the D scale under the hairline. Since there are two groups in the original number, there are two digits to the left of the decimal point in the root.

Thus, $\sqrt{4,521} = 67.2$

Example: $\sqrt{0.000741}\sqrt{0.00'07'41}$

(First figure is zero in this group)

Place the hairline over 741 on the left half of the A scale. Read the digit sequence of the root, 272, under the hairline on the D scale. Since there is one pair of zeros to the left of the group containing the first digit, there is one zero between the decimal point and the first significant digit of the root. Thus,

$$\sqrt{0.000741} = 0.0272$$

Practice problems: Evaluate each of the following by means of a slide rule:

1. $(17.75)^2$
2. $(0.65)^2$
3. $\sqrt{9.42}$
4. $\sqrt{0.074}$

Answers:

1. 315
2. 0.422
3. 3.07
4. 0.272

CUBES AND CUBE ROOTS

Cubes and cube roots are read on the K and D scales of the slide rule. On the K scale are compressed three complete logarithmic scales in the same space as that of the D scale. Thus, any logarithm on the K scale is three times the logarithm opposite it on the D scale. To cube a number by logarithms, we multiply its logarithm by three. Therefore, the logarithms of cubed numbers will lie on the K scale opposite the numbers on the D scale.

As with the other slide rule scales mentioned, the numbers the logarithms represent, rather than the logarithmic notations, are printed on the rule. In the left-hand third of the K scale, the numbers range from 1 to 10; in the middle third they range from 10 to 100; and in the right-hand third, they range from 100 to 1,000. To cube a number, find the number on the D scale, place the hairline over it, and read the digit sequence of the cubed number on the K scale under the hairline.

Placing the Decimal Point

The decimal point of a cubed whole or mixed number may be easily placed by application of the following rules:

1. If the cubed number is located in the left third of the K scale, its number of digits to the left of the decimal point is 3 times the number of digits to the left of the decimal point in the original number, less 2.
2. If the cubed number is located in the middle third of the K scale, its number of digits is 3 times the number of digits of the original number, less 1.
3. If the cubed number is located in the right third of the K scale, its number of digits is 3 times the number of digits of the original number.

Example: $(1.6)^3$

Solution: Place the hairline over 16 on D scale. Read the digit sequence, 409, on the K scale under the hairline. Number of digits to left of decimal point in the number 1.6 is 1 and the cubed number is in the left-hand third of the K scale.

$$3 \times (\text{No. of digits}) - 2 = (3 \times 1) - 2 = 1$$

Therefore,

$$(1.6)^3 = 4.09$$

Example: $(4.1)^3$

Digit sequence = 689.

Solution: Number of digits to left of decimal point in the number 4.1 is 1, and the cubed number is in the middle third of the K scale.

$$3 \times (\text{No. of digits}) - 1 = (3 \times 1) - 1 = 2$$

Therefore,

$$(4.1)^3 = 68.9$$

Example: $(52)^3$

Solution: Digit sequence = 141.

Number of digits to left of decimal point in the number 52 is 2, and the cubed number is in the right-hand third of the K scale.

$$3 \times \text{No. of digits} = 3 \times 2 = 6$$

Therefore,

$$(52)^3 = 141{,}000$$

Positive Numbers Less Than One

If positive numbers less than one are to be cubed, count the zeros between the decimal point and the first nonzero digit. Consider the count negative. Then the number of zeros between the decimal point and the first significant digit of the cubed number may be found as follows:

1. Left third of K scale: Multiply the zeros counted by 3 and subtract 2.
2. Middle third of K scale: Multiply the zeros counted by 3 and subtract 1.
3. Right third of K scale: Multiply the zeros counted by 3.

Example: Cube 0.034

Solution: Digit sequence = 393

Zero count of 0.034 = –1, and 393 is in the middle third of the K scale.

$$3 \times (\text{No. of zeros}) - 1 = (3 \times -1)-1 = -4$$

Therefore,

$$(0.034)^3 = 0.0000393$$

Practice problems: Cube the following numbers using the slide rule.

1. 21
2. 0.7
3. 0.0128
4. 404

Answers:

1. 9260
2. 0.342
3. 0.0000021
4. 66,000,000

Cube Roots

Taking the cube root of a number on the slide rule is the inverse process of cubing a number. To take the cube root of a number, find the number on the K scale, set the hairline over it, and read the cube root on the D scale under the hairline.

POSITIONING NUMBERS ON THE K SCALE

Since a given number can be located in three positions on the K scale, the question arises as to which third of the K scale to use when locating a number. Generally, the left index, the left middle index, the right middle index, and the right index are considered to be numbered as shown in figure.

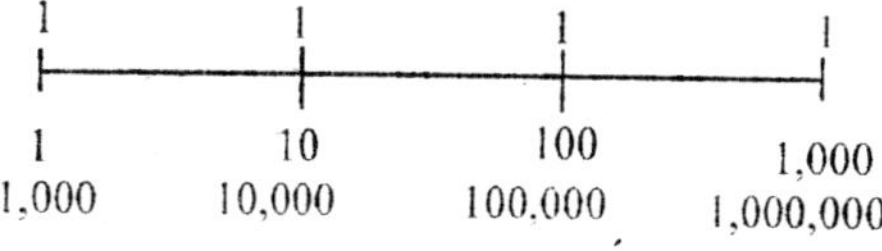

Fig. Powers of 10 Associated with K-scale indices.

A system similar to that used with square roots may be used to locate the position of a number on the K scale. Groups of three are used rather than groups of two. The grouping for cube root is illustrated as follows:

1. $\sqrt[3]{40,531.6}$, $\sqrt[3]{40'531.6}$
2. $\sqrt[3]{4,581.43}$, $\sqrt[3]{4'561.43}$
3. $\sqrt[3]{0.000043}$, $\sqrt[3]{0.000'043}$

For whole or mixed numbers the following rules apply:

1. If the left-hand group contains one digit, locate the number in the left third of the K scale.
2. If the left group contains two digits, locate the number in the middle third of the K scale.
3. If the left group contains three digits, locate the number in the right third of the K scale.

The following examples illustrate the foregoing rules:

1. $\sqrt[3]{4'561.43}$

   (One digit)-left third K scale.

2. $\sqrt[3]{40'531.6}$

   (Two digits)-middle third K scale.

3. $\sqrt[3]{453'361}$

   (Three digits)-right third of K scale.

For positive numbers less than one, look from left to right and find the first group that contains a digit other than zero.

1. If the first two figures in this group are zeros, locate the number in the left third of the K scale.
2. If only the first figure in this group is zero, locate the number in the middle third of the K scale.
3. If the first figure of the group is not zero, locate the number in the right third of the K scale.

The following examples illustrate these rules:

1. $\sqrt[3]{0.000'004'53}$

(Two zeros)-left third K scale.

2. $\sqrt[3]{0.000'050'43}$

(One zero)-middle third K scale.

3. $\sqrt[3]{0.000'000'430}$

(No zero)-right third K scale.

PLACING THE DECIMAL POINT

To place the decimal point in the cube root of a number, we use the system of marking off in groups of three as shown above.

For whole or mixed numbers, there is one digit in the root to the left of the decimal point for every group marked in the original number. Thus,

$$\sqrt[3]{4'531.6}$$

(Two digits in root to left of decimal point.)

For positive numbers less than one, there will be one zero in the root between the decimal point and the first significant digit for every three zeros counted between the decimal point and the first significant digit of the original number. Thus,

$$\sqrt[3]{0.000'000'004}$$

(Two zeros between decimal point and first significant digit of root.)

Example: $\sqrt[3]{216000.4}$, $\sqrt[3]{216'000.4}$

(Three digits in left group)

Place the hairline over 216 in the right third of the K scale. Read the digit sequence, 6, under the hairline on the D scale.

Since there are two groups in the original number, there are two digits to the left of the decimal point in the root. Thus,

$$\sqrt[3]{216000.4} = 60$$

Example: $\sqrt[3]{0.0000451} = \sqrt[3]{0.000'045'1}$

(Only first figure is zero in this group)

Place the hairline over 451 in the middle third of the K scale. Read the digit sequence, 357, under the hairline on the D scale. Since there is one group of three zeros, there is one zero between the decimal point and the first significant digit of the root. Thus,

$$\sqrt[3]{0.0000451} = 0.0357$$

POWERS OF 10

To take the cube root of a power of 10, mark it off as explained in the preceding paragraphs. The number in the left group will then be 1, 10, or 100. We know that the cube root of 10 is a number between 2 and 3. Thus, for the cube root of any number whose left group is 10, use the K scale index which lies between 2 and 3 on the D scale. The cube root of 100 lies between 4 and 5. Therefore, for a number whose left group is 100, use the K scale index between 4 and 5 on the D scale.

Practice problems: Following are some problems and the digit sequence (d. 6.) of the roots. Locate the decimal point for each root.

1. $\sqrt[3]{0.000023}$ d. s. 2844
2. $\sqrt[3]{0.051}$ d. s. 371
3. $\sqrt[3]{127}$ d. s. 5026
4. $\sqrt[3]{204,000}$ d. s. 589
5. $\sqrt[3]{734,000,000}$ d. s. 902
6. $\sqrt[3]{4.913}$ d. s. 17

Answers:

1. 0.02844
2. 0.371
3. 5.026
4. 58.9
5. 902
6. 17

Chapter 5

Fundamentals of Algebra

The numbers and operating rules of arithmetic form a part of a very important branch of mathematics called ALGEBRA. Algebra extends the concepts of arithmetic so that it is possible to generalize the rules for operating with numbers and use these rules in manipulating symbols other than numbers. It does not involve an abrupt change into a distinctly new field, but rather provides a smooth transition into many branches of mathematics with a continuation of knowledge already gained in basic arithmetic.

The idea of expressing quantities in a general way, rather than in the specific terms of arithmetic, is fairly common.

A typical example is the formula for the perimeter of a rectangle, $P = 2L + 2W$, in which the letter P represents perimeter, L represents length, and W represents width. It should be understood that $2L = 2(L)$ and $2W = 2(W)$. If the L and the W were numbers, parentheses or some other multiplication sign would be necessary, but the meaning of a term such as 2L is clear without additional signs or symbols.

All formulas are algebraic expressions, although they are not always identified as such. The letters used in algebraic expressions are often referred to as Literal Numbers (literal implies "letteral"). Another typical use of literal numbers is in the statement of mathematical laws of operation.

COMMUTATIVE LAWS

Remember that the commutative laws refer to those situations in which the factors and terms of an expression are rearranged in a different order.

ADDITION

The algebraic form of the commutative law for addition is as follows:

$$a + b = b + a$$

From this law, it follows that

$$a + (b + c) = a + (c + b) = (c + b) + a$$

In words, this law states that the sum of two or more addends is the same regardless of the order in which the addends are arranged.

The arithmetic example in earlier chapter shows only one specific numerical combination in which the law holds true. In the algebraic example, *a, b,* and *c* represent any numbers we choose, thus giving a broad inclusive example of the rule. (Note that once a value is selected for a literal number, that value remains the same wherever the letter appears in that particular example or problem. Thus, if we give a the value of 12, in the example just given, as value is 12 wherever it appears.)

MULTIPLICATION

The algebraic form of the Commutative law for multiplication is as follows:

$$ab = ba$$

In words, this law states that the product of two or more factors is the same regardless of the order in which the factors are arranged.

ASSOCIATIVE LAWS

The associative laws of addition and multiplication refer to the grouping (association) of terms and factors in a mathematical expression.

ADDITION

The algebraic form of the associative law for addition is as follows:

$$a + b + c = (a + b) + c = a + (b + c)$$

In words, this law states that the sum of three or more addends is the same regardless of the manner in which the addends are grouped.

MULTIPLICATION

The algebraic form of the associative law for multiplication is as follows:

$$a \bullet b \bullet c = (a \bullet b) \bullet c = a \bullet (b \bullet c)$$

In words, this law states that the product of three or more factors is the same regardless of the manner in which the factors are grouped.

DISTRIBUTIVE LAW

The distributive law refers to the distribution of factors among the terms of an additive expression. The algebraic form of this law is as follows:

$$a(b + c) = ab + ac$$

From this law, it follows that: If the sum of two or more quantities is multiplied by a third quantity, tine product is found by applying the multiplier to each of the original quantities separately and summing the resulting expressions.

ALGEBRAIC SUMS

The word "sum" has been used several times in this discussion, and it is important to realize the full implication where algebra is concerned. Since a literal number may represent either a positive or a negative quantity, a sum of several literal numbers is always understood to be an Algebraic Sum. That is, it is the sum that results when the algebraic signs of all the addends are taken into consideration. The following problems illustrate the procedure for finding an algebraic sum:

Let $a = 3, b = -2$, and $c = 4$.

Then $a + b + c = (3) + (-2) + (4) = 5$

Also, $a - b - c = a + (-b) + (-c) = 3 + (+2) + (-4) = 1$

The second problem shows that every expression containing two or more terms to be combined by addition and subtraction may be rewritten as an algebraic sum, all negative signs being considered as belonging to specific terms and all operational signs being positive. It should be noted, in relation to this subject, that the laws of signs for algebra are the same as those for arithmetic.

ALGEBRAIC EXPRESSIONS

An algebraic expression is made up of the signs and symbols of algebra. These symbols include the Arabic numerals, literal numbers, the signs of operation, and so forth. Such an expression represents one number or one quantity. Thus, just as the sum of 4 and 2 is one quantity, that is, 6, the sum of c and d is one quantity, that is, $c + d$.

Likewise, $\frac{a}{b}, \sqrt{b}\, ab,\ a - b$, and so forth, are algebraic expressions each of which represents one quantity or number.

Longer expressions may be formed by combinations of the various signs of operation and the other algebraic symbols, but no matter how complex such expressions are they still represent one number. Thus the algebraic expression $\frac{-a + \sqrt{2a + b}}{6} - c$ is one number

The arithmetic value of any algebraic expression depends on the values assigned to the literal numbers. For example, in the expression $2x^2 - 3ay$, if $x = -3$, $a = 5$, and $y = 1$, then we have the following:

$$2x^2 - 3ay = 2(-3)^2 - 3(5)(1) = 2(9) - 15 = 18 - 15 = 3$$

Notice that the exponent is an expression such as $2x^2$ applies only to the x. If it is desired to indicate the square of $2x$, rather than 2 times the square of x, then parentheses are used and the expression becomes $(2x)^2$.

Practice problems: Evaluate the following algebraic expressions when $a = 4$, $b = 2$, $c = 3$, $x = 7$, and $y = 5$. Remember, the order of operation is multiplication, division, addition, and subtraction.

1. $3x + 7y - c$
2. $xy - 4a^2$
3. $\frac{ax}{b} + y$
4. $c + \frac{ay^2}{b}$

Answers:

1. 53
2. –29
3. 19
4. 53

TERMS AND COEFFICIENTS

The terms of an algebraic expression are the parts of the expression that are connected by plus and minus signs. In the expression $3abx + cy - k$, for example, $3abx$, cy, and k are the term's of the expression.

An expression containing only one term, such as $3ab$, is called a monomial (mono means one). A binomial contains two terms; for example, $2r + by$. A trinomial consists of three terms. Any expression containing two or more terms may also be called by the general name, polynomial (poly means many).

Usually special names are not given to polynomials of more than three times. The expression $x^3 - 3x^2 + 7x + 1$ is a polynomial of four terms. The trinomial $x^2 + 2x + 1$ is an example of a polynomial which has a special name.

Practice problems: Identify each of the following expressions as a monomial, binomial, trinomial, or polynomial. (Some expressions may have two names.)

1. x
2. $3y + a + b$
3. abx
4. $4 + 2b + y + z$
5. $3y^2 + 4$
6. $\frac{2y}{6} + 1$

Answers:

1. Monomial
2. Trinomial (also polynomial)
3. Monomial
4. Polynomial
5. Binomial (also polynomial)
6. Binomial (also polynomial)

In general, a Coefficient of a term is any factor or group of factors of a term by which the remainder of the term is to be multiplied. Thus in the term $2axy$, $2ax$ is the coefficient of y, $2a$ is the coefficient of xy, and 2 is the coefficient of axy. The word "coefficient" is usually used in reference to that factor which is expressed in Arabic numerals. This factor is sometimes called the Nunerical Coefficient. The numerical coefficient is customarily written as the first factor of the term. In $4x$, 4 is the numerical coefficient, or simply the coefficient, of x. Likewise, in $24xy^2$, 24 is the coefficient of xy^2 and in $16(a + b)$, 16 is the coefficient of $(a + b)$. When no numerical coefficient is written it is understood to be 1. Thus in the term xy, the coefficient is 1.

COMBINING TERMS

When arithmetic numbers are connected by plus and minus signs, they can always be combined into one number. Thus,

$$5-7\frac{1}{2}+8=5\frac{1}{2}$$

Here three numbers are added algebraically (with due regard for sign) to give one number. The terms have been combined into one term. Terms containing literal numbers can be combined only if their literal parts are the same.

Terms containing literal factors in which the same letters are raised to the same power are called like terms. For example, $3y$ and $2y$ are like terms since the literal parts are the same. Like terms are added by adding the coefficients of the like parts. Thus, $3y + 2y = 5y$ just as 3 bolts + 2 bolts = 5 bolts. Also $3a^2b$ and a^2b are like; $3a^2b + a^2b = 4a^2b$ and $3a^2b - a^2b = 2a^2b$. The numbers ay and by are like terms with respect to y. Their sum could be indicated in two ways: $ay + by$ or $(a + b)y$. The latter may be explained by comparing the terms to denominate numbers. For instance, a bolts + b bolts = $(a + b)$ bolts.

Like terms are added or subtracted by adding or subtracting the numerical coefficients and placing the result in front of the literal factor, as in the following example

$$7x^2 - 5x^2 = (7 - 5)x^2 = 2x^2$$

$$5b^2x - 3ay^2 - 8b^2x + 10ay^2 = -3b^2x + 7ay^2$$

Dissimilar or unlike terms in an algebraic expression cannot be combined when numerical values have not been assigned to the literal factors.

For example, $-5x^2 + 3xy - 8y^2$ contains three dissimilar terms. This expression cannot be further simplified by combining terms through addition or subtraction. The expression may be rearranged as $x(3y - 5x) - 8y^2$ or $y(3x - 8y) - 5x^2$, but such a rearrangement is not actually a simplification.

Practice problems: Combine like terms in the following expression:

1. $2a + 4a$
2. $y + y^2 + 2y$
3. $4\frac{ay}{c} - \frac{ay}{c}$

4. $2ay^2 - ay^2$
5. $bx^2 + 2bx^2$
6. $2y + y^2$

Answers:

1. $6a$
2. $y^2 + 3y$
3. $3\frac{ay}{c}$
4. $4ay^2$
5. $3bx^2$
6. $2y + y^2$

SYMBOLS OF GROUPING

Often it is desired to group two or more terms to indicate that they are to be considered and treated as though they were one term even though there may be plus and minus signs between them. The symbols of grouping are parentheses () (which we have already used), brackets [ ], braces { }, and the vinculum.

The vinculum is sometimes called the "overscore." The fact that $-7 + 2 - 5$ is to be subtracted from 15, for example, could be indicated in any one of the following ways:

$$15-(-7+2-5)$$
$$15-[-7+2-5]$$
$$15-\{-7+2-5\}$$
$$15-\overline{-7+2-5}$$

Actually the vinculum is seldom used except in connection with a radical sign, such as in $\sqrt{a+b}$ or in a Boolean algebra expression. Boolean algebra is a specialized kind of symbolic notation which is discussed in Mathematics, Volume 3, NavPers 10073.

Parentheses are the most frequently used symbols of grouping. When several symbols are needed to avoid confusion in grouping, parentheses usually comprise the innermost symbols, followed by brackets, and then by braces as the outermost symbols. This arrangement of grouping symbols is illustrated as follows:

$$2x - \{3y + [-8 - 5y - (x - 4)]\}$$

REMOVING AND INSERTING GROUPING SYMBOLS

Discussed in the following paragraphs are various rules governing the removal and ineertion of parentheses, brackets, braces, and the vinculum. Since the rules are the same for all grouping symbols, the discussion in terms of parentheses will serve as a basis for all.

Removing Parentheses

If parentheses are preceded by a minus sign, the entire quantity enclosed must be regarded as a subtrahend. This means that each term of the quantity in parentheses is subtracted from the expression preceding the minus sign. Accordingly, parentheses preceded by a minus sign can be removed, if the signs of all terms within the parentheses are changed.

This may be explained with an arithmetic example. We recall that to subtract one number from another, we change the sign of the subtrahend and proceed as in addition. To subtract –7 from 16, we change the sign of –7 and proceed as in addition, as follows:

$$16 - (-7) = 16 + 7 = 23$$

It is sometimes easier to see the result of changing signs in the subtrahend if the minus sign preceding the parentheses is regarded as a multiplier. Thus, the thought process in removing parentheses from an expression such as – (4 – 3 + 2) would be as follows: Minus times plus is 'minus, so the first term of the expression with parentheses removed is – 4. (Remember that the 4 in the original expression is understood to be a + 4, since it has no sign showing.) Minus times minus is plus, so the second term is +3. Minus times plus is minus, so the third term is –2. The result is – 4 + 3 – 2, which reduces to –3.

This same result can be reached just as easily, in an arithmetic expression, by combining the numbers within the parentheses before applying the negative sign which precedes the parentheses. However, in an algebraic expression with no like terms such combination is not possible. The following example shows how the rule for removal of parentheses is applied to algebraic expressions:

$$2a - (-4x + 3by) = 2a + 4x - 3by$$

Parenthesis preceded by a plus sign can be removed without any other changes, m the following example shows:

$$2b + (a - b) = 2b + a - b - a + b$$

Many expressions contain more than one set of parentheses, brackets, and other symbols of grouping. In removing symbols of grouping, it is possible to proceed from the outside inward or from the inside outward. For the beginner, it is simpler to start on the inside and work toward the outside, collecting terms and simplifying as one proceeds. In the following example the inner grouping symbols are removed first:

$$\begin{aligned} & 2a - [x + (x - 3a) - (9a - 5x)] \\ & = 2a - [x + x - 3a - 9a + 5x] \\ & = 2a - [7x - 12a] \\ & = 14a - 7x \end{aligned}$$

Enclosing Terms in Parentheses

When it is desired to enclose a group of terms in parentheses, the group of terms remains unchanged if the sign preceding the parentheses is positive. This is illustrated as follows:

$$3x - 2y + 7x - y = (3x - 2y) + (7x - y)$$

Note that this agrees with the rule for removing parentheses preceded by a plus sign.

If terms are enclosed within parentheses preceded by a minus sign, the signs of all the terms enclosed must be changed as in the following example:

$$3x - 2y + 7x - y = 3x - (2y - 7x + y)$$

Practice problems: In problems 1.through 4, remove the symbols of grouping and combine like terms. In problems 5 through 8, enclose the first two terms in parentheses preceded by a plus sign (understood) and the last two in parentheses preceded by a minus sign.

1. $6a - (4a - 3)$
2. $3x + [2x - 4y(6 - 4x)] + 2y - (3 - x + 3y)$
3. $-a + [-a - (2a + 3)] + 3$
4. $(4x - 3ay) - (4a - b) + 16$
5. $4a - 3b - 2c + 4d$
6. $-2 - 3x + 4y - z$
7. $x + 4y + 3z + 7$
8. $-4 + 2a - 6c + 3d$

Andwers:

1. $2a + 3$
2. $6x + 16xy - 25y - 3$
3. $-4a$
4. $7x - 3ay - 4a + b + 16$
5. $(4a - 3b) - (2c - 4d)$
6. $(-2 - 3x) - (-4 + z)$
7. $(x + 4y) - (-3z - 7)$
8. $(-4 + 2a) - (6c - 3d)$

EXPONENTS AND RADICALS

Exponents and radicals have the same meaning in algebra as they do in arithmetic. Thus, if n represents any number then $n^2 = n \cdot n$, $n^3 = n \cdot n \cdot n$, etc. By the same reasoning, n^m means that

n is to be taken as a factor m times. That is, n^m is equal to $n \cdot n \cdot n$.....with n appearing m times. The series of dots, called ellipsis (not to be confused with the geometric figure having a similar name, ellipse), represents continuation of the same pattern or the same symbol.

The rules of operation with exponents are also the same in algebra as in arithmetic. For example, $n^2 \cdot$. Some care is exponents is a number or a letter. Thus the general form can be expressed as follows:

$$n^a \cdot n^b = n^{a+b}$$

In words, the general rule for multiplication involving exponents is as follows: When multiplying terms whose literal factors are like, the exponents are added. This rule may be applied to problems involving division, if all expressions containing exponents in denominators are rewritten as expressions with negative exponents. For example, the fraction $\frac{x^2y}{xy^2}$ can be rewritten as $(x^2y)(x^{-1}y^{-2})$, which is equal to $(x^{2-1})(y^{1-2})$. This reduces to xy^{-1}, or x/y Notice that the result is the same as it would have been if we had simply subtracted the exponents of literal factors in the denominator from the exponents of the same literal factors in the numerator.

The algebraic rules for radicals also remain the same as those of arithmetic. In arithmetic, $\sqrt{4} = 4^{1/2} = 2$. Likewise, in algebra $\sqrt{a} = a^{1/2}$ and $\sqrt[2]{a} = a^{1/n}$.

MULTIPLYING MONOMIALS

If a monomial such as 3abc is to be multiplied by a numerical multiplier, for the coefficient alone is multiplied, following example:

$$5 \times 3abc = 15abc$$

When the numerical factor is not the initial factor of the expression, as in $x(2a)$, the result of the multiplication is not written as x^2a. Instead, the numerical factor is interchanged with literal factors by use of the commutative law of multiplication. The literal factors are usually interchanged to place them in alphabetical order, and the final result is as follows:

$$x(2a) = 2ax$$

The rule for multiplication of monomials may be stated as follows: Multiply the numerical coefficients to form the coefficient of the product. Multiply the literal factors, combining exponents in like factors, to form the literal part of the product. The complete process is illustrated in the following example:

$$(2ab)(3a^2)(2b^3) = 12a^{1+2}b^{1+3} = 12a^3 b^4$$

Practice problems: Perform the indicated operations:

Answers:

1. $(2x^2)\ (5x^5)$
2. $(-5ab^2)(2a^2b)$
3. $(-4x^4y)(-3xy^4)$
4. $(2^a)(2^b)$
5. $(-4a^3)^2$
6. $(3a^2b)^2$

Answers:

1. $10x^7$
2. $-10a^3b^3$
3. $12x^5y$
4. 2^{n+b}
5. $16a^6$
6. $9a^4b^2$

DIVIDING MONOMIALS

As may be expected, the process of dividing is the inverse of multiplying. Because $3 \times 2a = 6a$, $6a \div 3 = 2a$, or $6a \div 2 = 3a$. Thus, when the divisor is numerical, divide the coefficient of the dividend by the divisor.

When the divisor contains literal parts that are also in the dividend, cancellation may be performed as in arithmetic. For example, $6ab \div 3a$ may be written as follows:

$$\frac{(2)(3a)(b)}{3a}$$

Cancellation of the common literal factor, $3a$, from the numerator and denominator leaves $2b$ as the answer for this division problem.

When the same literal factors appear in both the divisor and the dividend, but with different exponents, cancellation may still be used, as follows:

$$\frac{14a^3b^3x}{-21a^2b^5x} = \frac{(7)(2)a^2ab^3x}{(7)(-3)a^2b^3b^2x} = \frac{2a}{-3b^2} = -\frac{2a}{3b}2$$

This same problem may be solved without thinking in terms of cancellation, by rewriting with negative exponents as follows:

$$\frac{14a^3b^3x}{-21a^2b^5x} = \frac{2a^{3-2}b^{3-5}x^{1-1}}{-3} - \frac{2ab^{-2}}{-3} = \frac{2a}{-3b^2} = -\frac{2a}{3b^2}$$

Practice problems: Perform the indicated operations:

1. $\frac{x^5}{x^6}$
2. $\frac{a^9b^4}{a^6b^3}$
3. $\frac{a^2bc^2}{abc}$
4. $\frac{a^2b}{ab^2}$
5. $\sqrt{16x^4y^6}$
6. $\sqrt{x^4ay2a}$
7. $\frac{5a^4b}{10a^2b^3}$

8. $\dfrac{10x^2y^3z^4}{-5xy^2z^3}$

9. $\sqrt{a^6b^{6n}}$

10. $\sqrt{a^6b^{6n}}$

Anwers:

1. $x - 1$
2. $a^3 b$
3. ac
4. $\dfrac{a}{b}$
5. $\pm 4x^2y^3$
6. $\pm x^{2a} y^a$
7. $\dfrac{a^2}{2b^2}$
8. $-2xyz$
9. $\pm 10a^4 b^2$
10. $\pm a^3 y^{3n}$

OPERATIONS WITH POLYNOMIALS

Adding and subtracting polynomials is simply the adding and subtracting of their like terms. There is a great similarity between the operations with polynomials and denominate numbers. Compare the following examples:

1. Add 5 qt and 1 pt to 3 qt and 2 pt.

$$\begin{array}{r} 3qt + 2pt \\ \underline{3qt + 1pt} \\ 8qt + 3pt \end{array}$$

2. Add $5x + y$ to $3x + 2y$.

$$\begin{array}{r} 3x+2y \\ 3x+y \\ \hline 8x+3y \end{array}$$

One method of adding polynomials (shown in the above examples) is to place like terms in columns and to find the algebraic sum of the like terms. For example, to add $3a + b - 3c$, $3b + c - d$, and $2a + 4d$, we would arrange the polynomials as follows:

$$\begin{array}{l} 3a+b-3c \\ \quad\; 3b+\;c\;-\;d \\ 2a \qquad\qquad\; +4d \\ \hline 5a+4b-2c+3d \end{array}$$

Subtraction may be performed by using the same arrangement-that is, by placing terms of the subtrahend under the like terms of the minuend and carrying out the subtraction with due regard for sign.

Remember, in subtraction the signs of all the terms of the subtrahend must first be mentally changed and then the process completed as in addition. For example, subtract $10a + b$ from $8a - 2b$, as follows:

$$\begin{array}{r} 8a-2b \\ 10a+b \\ \hline -2a-3b \end{array}$$

Again, note the similarity between this type of subtraction and the subtraction of denominate numbers.

Addition and subtraction of polynomials also can be indicated with the aid of symbols of grouping. The rule regarding changes of sign when removing parentheses preceded by a minus sign automatically takes care of subtraction.

For example, to subtract $10a + b$ from $8a - 2b$, we can use the following arrangement:

$$(8a - 2b)-(10a + b) = 8a -2b - 10a -b = -2a-3b$$

Similarly, to add $-3x + 2y$ to $-4x -5y$, we can write

Practice problems: Add as indicated, in each of the following problems:

1. $3a + b, 2a + 5b$
2. $(6a^3t + 3a^2t + st + 5) + (a^3t - 5)$
3. $4a + b + c, a + c - d, Sa + 2b + 2c$
4. $4x + 2y,\ \begin{array}{l} 3x - y + z \\ \underline{x \qquad\quad -z} \end{array}$

In problems 5 through 8, perform the indicated operations and combine like terms.

5. $(2a + b) - (3a + 5b)$
6. $(5x^3y + 3x^2y) - (x^3y)$
7. $(x + 6) + (3x + 7)$
8. $(4a^2 - b) - (2a^2 + b)$

Answers:

1. $5a + 6b$
2. $7s^3t + Ss^2t + Bt$
3. $8a + 3b + 4c - d$
4. $8x + y$
5. $-(a + 4b)$
6. $4x + 13$
7. $4x + 13$
8. $2(a^2 - b)$

MULTIPLICATION OF A POLYNOMIAL BY A MONOMIAL

We can explain the multiplication of a polynomial by a monomial by using an arithmetic example. Let it be required to multiply the binomial expression, 7 – 2, by 4. We may write this 4 × (7 – 2)orsimply 4(7 – 2). Now 7 – 2 = 5. Therefore, 4(7 – 2) = 4(5) = 20. Now, let us then subtract. Thus, 4(7 – 2) = (4 × 7) – (4 × 2) = 20. Both methods give the same result. The second method makes use of the distributive law of multiplication.

When there are literal parts in the expression to be multiplied, the first method cannot be used and the distributive method must be employed. This is illustrated in the following examples:

$$4(5 + a) = 20 + 4a$$

$$3(a + b) = 3a + 3b$$

$$ab(x + y - z) = abx + aby - abz$$

Thus, to multiply a polynomial by a monomial, multiply each term of the polynomial by the monomial.

Practice problems: Multiply as indicated:

1. $2a(a - b)$
2. $4a^2 (4a^2 + 5a + 2)$
3. $-4x(-y-3z)$
4. $(2a^3(a^2 - ab)$

Answers:

1. $2a^2 - 2ab$
2. $4a^4\ 20a^3 + 8a^2$
3. $4xy(12xy)$
4. $2a^5 - 2a^4b$

MULTIPLICATION OF A POLYNOMIAL BY A POLYNOMIAL

As with the monomial multiplier, we explain multiplication of a polynomial by a polynomial by use of an arithmetic example. To multiply (3 + 2)(6 – 4), we could do the operation within the parentheses first and then multiply, as follows:

$$(3 + 2)(6 - 4) = (5)(2) = 10$$

However, thinking of the quantity (3 + 2) as one term, we can use the method described for a (3 + 2), with the following result:

$$(3 + 2)(6 - 4) = [(3 + 2) \times 6 - (3 + 2) \times 4]$$

Now considering each of the two resulting products

separately, we note that each is a binomial multiplied by a monomial.

The first is

$$(3 + 2)6 = (3 \times 6) + (2 \times 6)$$

and the second is

$$-(3 + 2)4 = -[(3 \times 4) + (2 \times 4)] = -(3 \times 4) - (2 \times 4)$$

Thus we have the following result:

$$(3 + 2)(6 - 4) = (3 \times 6) + (2 \times 6) - (3 \times 4) - (2 \times 4)$$
$$= 18 + 12 - 12 - 8 = 10$$

The complete product is formed by multiplying each term of the multiplicand separately by each term of the multiplier and combining the results with due regard to signs.

Now let us apply this method in two examples involving literal numbers.

1. $(a + b)(m + n) = am + an + bm + bn$
2. $(2b + c)(r + s + 3t - u) = 2br + 2bs + 6bt - 2bu + cr + cs + 3ct - cu$

The rule governing these examples is stated as follows: The product of any two polynomials is found by multiplying each term of one by each term of the other and adding the results algebraically.

It is often convenient, especially when either of the expressions contains more than two terms, to place the polynomial with the fewer terms beneath the other polynomial and multiply term by term beginning at the left. Like $3x^2 - 7x - 9$ and $2x - 3$. The procedure is

$$\begin{array}{l} 3x^2 - 7x - 9 \\ 2x - 3 \\ \hline 6x^3 - 14x^2 - 18x \\ \quad -9x^2 + 21x + 27 \end{array}$$

Practice problems: In the following problems, multiply and combine like terms:

1. $(2a - 3)(a + 2)$

2. $(ax + b)(ax - b)$

3. $\dfrac{x^3 + 5x^2 - x + 2}{2x + 3}$

4. $\dfrac{2a^2 - 5x^2 - b^2}{a + b}$

ANSWERS

1. $2a^2 + a - 6$
2. $a^2x^2 - b^2$
3. $2x^4 + 13x^3 + 13x^2 + x + 6$
4. $2x^3 + 7a^2b + 4ab^2 - b^3$

SPECIAL PRODUCTST

The products of certain binomials occur frequently. It is convenient to remember the form of these products so that they can be written immediately without performing the complete multiplication process. We present four such special products as follows, and then show how each is derived:

1. Product of the sum and difference of two numbers.

EXAMPLE: $(x - y)(x + y) = x^2 - y^2$

2. Square the sum of two numbers.

EXAMPLE: $(x + y)^2 = x^2 + 2xy + y^2$

3. Square of the difference of two numbers.

EXAMPLE: $(x - y)^2 = x^2 - 2xy + y^2$

4. Product of two binomials having a common term.

EXAMPLE: $(x + a)(x + b) = x^2 + (a + b)x + ab$

Product of Sum and Difference

The product of the sum and difference of two numbers is equal to the square of the first number minus the square of the second number. If, for example, $x - y$ is multiplied by $x + y$, the middle terms cancel one another. The result is the square of x minus the square of y, as shown in the following illustration:

$$\begin{array}{l} x \quad -y \\ x \;+\; y \\ \hline x^2 - xy \\ \quad\; + xy - y^2 \\ \hline x^2 \qquad\quad -y^2 \end{array}$$

By keeping this rule in mind, the product of the sum and difference of two numbers can be written down immediately by writing the difference of the squares of the numbers. For example, consider the following three problems:

$$(x + 3)\,(x - 3) = x^2 - 3^2 = x^2 - 9$$

$$(5a + 2b)\,(5a - 2b) = (5a)^2 - (2b)^2 = 25a^2 - 4b^2$$

$$(4x + 4y)\,(7x - 4y) = 49x^2 - 16y^2$$

RATIONALIZING DENOMINATORS.- The product of the sum and difference of two numbers is useful in rationalizing a denominator that is a binomial. For example, in a fraction such as $\frac{2}{\sqrt{2} = 6}$ the denominator can be altered so that no radical terms appear in it. (This process is called rationalizing.)

The denominator must be multiplied by $\sqrt{2}$ + 6, which is called the conjugate of $\sqrt{2}$ − 6. Since the value of the original fraction would be changed if we multiplied only the denominator, our multiplier must be applied to both the numerator and the denominator. Multiplying the original fraction by $\frac{\sqrt{2} - 6}{\sqrt{2} + 6}$ is, in effect, the same as multiplying it by 1. The result of rationalizing the denominator of this fraction is as follows:

$$\frac{2}{\sqrt{2} - 6} \cdot \frac{\sqrt{2} + 6}{\sqrt{2} + 6} = \frac{2(\sqrt{2} + 6)}{(\sqrt{2})^2 - 6^2} = \frac{2(\sqrt{2} + 6)}{2 - 36}$$

$$= \frac{2(\sqrt{2} + 6)}{2(1 - 18)} = \frac{2\sqrt{2} + 6)}{2(-17)}$$

$$= \frac{\sqrt{2} + 6}{-17}$$

MENTAL MULTIPLICATION

The product of the sum and difference can be utilized to mentally multiply two numbers that differ from a multiple of 10 by the same amount, one greater and the other less. For example, 67 is 3 less than 70 while 73 is 3 more than 70.

The product of 67 and 73 is then found as follows:

$$67(73) = (70 - 3)(70 + 3) = 70^2 - 3^2 = 4,900 - 9 = 4,891$$

Square of Sum or Difference

The square of the SUM of two numbers is equal to the square of the first number plus twice the product of the numbers plus the square of the second number. The square of the DIFFERENCE of the same two numbers has the same form, except that the sign of the middle term is negative.

These results are evident from multiplication. When x and y represent the two numbers, we obtain

$$\begin{array}{ll} x+y & x-y \\ \underline{x+y} & \underline{x-y} \\ x^2+xy & x^2-xy \\ \underline{\quad +xy+y^2} & \underline{\quad -xy-y^2} \\ x^2+2xy+y^2 & x^2-2xy-y^2 \end{array}$$

Applying this rule to the squares of the binomials $3a + 2b$ and $3a - 2b$, we have the following two cases:

1. $(3a+2b)^2 = 9a^2 - 12ab + 4b^2 = 9a^2 + 12ab + 4b^2$

   $= 9a^2 + 12ab + 4b^2$

2. $(3a+2b)^2 = 9a^2 - 12ab + 4b^2$

The square of the sum or difference of two numbers is applicable to squaring a binomial that contains one or two irrational terms, as in the following examples:

1. $(\sqrt{3}+8)^2 = (\sqrt{3})^2 + 2(8)(\sqrt{3}) + 64$

$$= 3 + 16\sqrt{3} + 64 = 67 + 16\sqrt{3}$$

2. $(\sqrt{3} - 8)^2 = (\sqrt{3})^2 - 2(8)(\sqrt{3}) + 64$

$$= 3 - 16\sqrt{3} + 64 = 67 - 16\sqrt{3}$$

3. $(\sqrt{5} - \sqrt{7})^2 = (\sqrt{5})^2 + 2\sqrt{5}\sqrt{7} + (\sqrt{7})^2$

$$= 5 + 2\sqrt{35} + 7 = 12 + 2\sqrt{35}$$

The square of the sum or difference of two numbers can be applied to the process of mentally squaring certain numbers. For example, 82^2 can be expressed as $(80 + 2)^2$ while 67^2 can be expressed as $(70 - 3)^2$. We find that

$$(80 + 2)^2 = 80^2 + 2(80)(2) + 2^2 = 6,400 + 320 + 4 = 6,724$$
$$(70 - 3)^2 = 70^2 - 2(70)(3) + 3^2 = 4,900 + 420 + 9 = 4,489$$

Binomials Having a Common Term

The binomials $x + 2$ and $x - 3$ have a common term, x. They have two unlike terms, +2 and –3. The product of these binomials is

$$\begin{array}{l} x + 2 \\ \underline{x - 3} \\ x^2 + 2x \\ \underline{\quad - 3x - 6} \\ x^2 - x - 6 \end{array}$$

Inspection of this product shows that it is obtained by squaring the common term, adding the sum of the unlike terms multiplied by the common term, and finally adding the product of the unlike terms.

Apply this rule to the product of $3y - 5$ and $3y + 4$. The common term is 3y; its square is $9y^2$. The sum of the unlike terms is $-5 + 4 = -1$; the sum of the unlike terms multiplied by the common term is $-3y$; and the product of the unlike terms is $-5(4) = -20$. The product of the two binomials is

$$(3y - 5)(3y + 4) = 9y^2 - 3y - 20$$

The product of two binomials having a common term is applicable to the multiplication of numbers like $\sqrt{3}+7$ and $\sqrt{3}-2$ which contain irrational terms. For example,

$$(\sqrt{3}+7)(\sqrt{3}-2)=(\sqrt{3})^2+5\sqrt{5}\sqrt{3}-14$$
$$=3+5\sqrt{3}-14$$

Practice problems: In problems 1 through 4, multiply and combine terms. In 5 through 8, simplify by using special products.

1. $(x+4)(x+2)$
2. $(\sqrt{a}-b)^2$
3. $(7a+4b)\,(7a-4b)$
4. $(ax+y)^2$
5. $\dfrac{2}{\sqrt{2}-2}$
6. 48(52)
7. $(\sqrt{3}+7)^2$
8. $(73)^2$

Answers:

1. x^2+6x+8
2. $a-2b\sqrt{a}+b^2$
3. $49a^2-16b^2$
4. $a^2x^2+2axy+y^2$
5. $-(\sqrt{2}+2)$
6. $(50-2)(50+2)=2496$
7. $52+14\sqrt{3}$
8. $(70+3)\,(70+3)=5329$

DIVISION OF A POLYNOMIAL BY A MONOMIAL

Division, like multiplication, may be distributive. Consider, for example, the problem (4 + 6 – 2) ÷ 2, which may be solved by adding the numbers within the parentheses and then dividing the total by 2. Thus,

$$\frac{4+6-2}{2}=\frac{8}{2}=4$$

Now notice that the problem may also-be solved distributively.

$$\frac{4+6-2}{2}=\frac{4}{2}+\frac{6}{2}=\frac{2}{2}=2+3-1=4$$

Caution: Do not confuse problems of the type just described with another type which is similar in appearance but not in final result. For example, in a problem such as 2 ÷ (4 + 6 – 2) the beginner is tempted to divide 2 successively by 4, then 6, and then –2, as follows:

$$\frac{2}{4\div6-2}\neq\frac{2}{4}+\frac{2}{6}-\frac{2}{2}$$

Notice that we have canceled the "equals" sign, because 2 + 8 is obviously not equal to 1/2 + 2/6. – 1. The distributive method applies only in those cases in which several different numerators are to be used with the same denominator. When literal numbers are present in an expression, the distributive method must be used, as in the following two problems: Quite often this division may be done mentally, and the intermediate steps need not be written out.

DIVISION OF A POLYNOMIAL BY A POLYNOMIAL

Division of one polynomial by another proceeds as follows:

1. Arrange both the dividend and the divisor in either descending or ascending powers of the same letter.
2. Divide the first term of the dividend by the first term of the divisor and write. the result as the first term of the quotient.

3. Multiply the complete divisor by the quotient just obtained, write the terms of the product under the like terms of the dividend, and subtract this expression from the dividend.
4. Consider the remainder as a new dividend and repeat steps 1, 2, and 3.

Example:

$$(10x^3 - 7x^2y - 16xyz^2 + 12y^3) + (5x - 6y)$$

Solution: In the example just shown, we began by dividing the first term, $10x^3$, of the dividend by the first term, $5x$, of the divisor.

The result is $2x^2$. This is the first term of the quotient. Next, we multiply the divisor by $2x^2$ and subtract this product from the dividend. Use the remainder as a new dividend. Get the second term, xy, in the quotient by dividing the first term, $5x^2y$, of the new dividend by the first term, $5x$,of the divisor. Multiply the divisor by xy and again subtract from the dividend. Continue the process until the remainder is zero or is of a degree lower than the divisor. In the example being considered, the remainder is zero (indicated by the double line at the bottom). The quotient is $2x^2 + xy - 2y^2$. The following long division problem is an example in which a remainder is produced:

$$\begin{array}{r|l}
 & x^2 - x + 3 \\
\hline
x + 3 & x^3 + 2x^2 \qquad\qquad +5 \\
 & \underline{x^3 + 3x^2} \\
 & \quad -x^2 \\
 & \quad \underline{-x^2 - 3x} \\
 & \qquad\qquad 3x + 5 \\
 & \qquad\qquad \underline{3x + 9} \\
 & \qquad\qquad\quad -4
\end{array}$$

The remainder is –4.

Notice that the term $-3x$ in the second step of this problem is subtracted from zero, since there is no term containing x in the dividend.

When writing down a dividend for long division, leave spaces for missing terms which may enter during the long division process. In arithmetic, division problems are often arranged as follows, in order to emphasize the relationship between the remainder and the divisor:

$$\frac{5}{2} = 2 + \frac{1}{2}$$

This same type of arrangement is used in algebra. For example, in the problem just shown, the results could be written as follows:

$$\frac{x^3 + 2x^2 + 5}{x+3} = x^2 - x + 3 = -\frac{4}{x+3}$$

Remember, before dividing polynomials arrange the terms in the dividend and divisor according to either descending or ascending powers of one of the literal numbers. When only one literal number occurs, the terms are usually arranged in order of descending powers..

For example, in the polynomial $2x^2 + 4x^3 + 5 - 7x$ the highest power among the literal terms is x^3. If the terms are arranged according to descending powers of x, the term in x^3 should appear first. The x^3 term should be followed by the x term, the x term, and finally the constant term. The polynomial arranged according to descending powers of x is $4x^3 + 2x^2 - 7x + 5$. Suppose that $4ab + b^2 + 15a^2$ is to be divided by $3a + 2b$. Since $3a$ can be divided evenly into $15a^2$, arrange the terms according to descending powers of a. The dividend takes the form $15a^2 + 4ab + b^2$.

Synthetic Division

Synthetic division is a shorthand method of dividing a polynomial by a binomial of the form $x - a$. For example, if $3x^4 + 2x^3 + 2x^2 - x - 6$ is to be divided by $x - 1$, the long form would be as follows: Notice that every alternate line of work in this example contains a term which duplicates the one above it.

Furthermore, when the subtraction is completed in each step, these duplicated terms cancel each other and thus have no effect on the final result.

Another unnecessary duplication results when terms from the dividend are brought down and rewritten prior to subtraction. By omitting these duplications, the work may be condensed as follows:

$$\begin{array}{r l} & 3x^3+5x^2\quad +7x+6 \\ x-1 & \overline{)\,3x^4+2x^3+2x^2-x-6} \\ & \quad -3x^3-5x^2-7x-6 \\ \hline & \quad\quad +5x^3+7x^2+6x\ 0 \end{array}$$

The coefficients of the dividend and the constant term of the divisor determine the results of each successive step of multiplication and subtraction. Therefore, we may condense still further by writing only the nonliteral factors, as follows:

$$\begin{array}{r l} & 3\quad +5+7+6 \\ -1 & \overline{)\,3+2+2-1-6} \\ & \quad -3-5-7-6 \\ \hline & 3\ +5+7+6\ \ 0 \end{array}$$

Notice that if the coefficient of the first term in the dividend is brought down to the last line, then the numbers in the last line are the same as the coefficients of the terms in the quotient. Thus we do not really need to write a separate line of coefficients to represent the quotient. Instead, we bring down the first coefficient of the dividend and make the subtraction "subtotals" serve as coefficients for the rest of the quotient, as follows:

$$\begin{array}{r|l} x-1 & 3\ \ 2\ \ 2\ -1\ -6 \\ & \ -3-5-7-6 \\ \hline 3 & \quad 5\ \ 7\ \ 6\ \ 0 \end{array}$$

The unnecessary writing of plus signs is also eliminated here. The use of synthetic division is limited to divisors of the form $x - a$, in which the degree of x is 1. Thus the degree of each term in the quotient is 1 less than the degree of the corresponding term in the dividend. The quotient in this example is as follows:

$$3x^3 + 5x^2 + 7x + 6$$

The sequence of operations in synthetic division may be summarized as follows, using as an example the division of $3x - 4x^2 + x^4 - 3$ by $x + 2$:

$$\begin{array}{r|rrrrr} & 1 & 0 & -4 & 3 & -3 \\ 2 & & 2 & -4 & 0 & 6 \\ \hline & 1 & -2 & 0 & 0 & -9 \end{array}$$

First, rearrange the terms of the dividend in descending powers of x. The dividend then becomes $x^4 - 4x^2 + 3x - 3$, with 1 understood as the coefficient of the first term. No x^3 term appears in the polynomial, but we supply a zero as a place holder for the x^3 position. Second, bring down the 1 and multiply it by the +2 of the divisor. Place the result under the zero, and subtract. Multiply the result (–2) by the +2 of the divisor, place the product under the –4 of the dividend, and subtract. Continue this process, finally obtaining $x^3 - 2x^2 + 3$ as the quotient. The remainder is –9.

Practice problems: In the following problems, perform the indicated operations. In 4, 5, and 6, first use synthetic division and then check your work by long division:

1. $(a^3 - 3a^2 + a) \div a$
2. $\dfrac{x^6 - 7x^5 + 4x^4}{x^2}$
3. $(10x^3 - 7x^2y - 16xy^2 + 12y^3) \div (2x^2 + xy - 2y^2)$
4. $(x^2 + 11x + 30) \div (x + 6)$
5. $(12 + x^2 - 7x) \div (x - 3)$
6. $(a^2 - 11a + 30) \div (a - 5)$.

Chapter 6

Factoring Polynomials

A factor of a quantity *N*, as defined in chapter 2 of this course, is any expression which can be divided into *N* without producing a remainder. Thus 2 and 3 are factors of 6, and the factors of $5x$ are 5 and x. Conversely, when all of the factors of *N* are multiplied together, the product is *N*. This definition is extended to include polynomials.

The factors of a polynomial are two or more expressions which, when multiplied together, give the polynomial as a product. For example, 3, x, and $x^2 - 4$ are factors of $3x^3 - 12x$, as the following equation shows:

$$(3)(x)(x^2 - 4) = 3x^3 - 12x$$

The factors 3 and x, which are common to both terms of the polynomial $3x^3 - 12x$, are called Common Factors. The distributive principle, mentioned in chapters 3 and 9 of this course, is an important part of the concept of factoring. It may be stated as follows:

If the sum of two or more quantities is multiplied by a third quantity, the. product is found by applying the multiplier to each of the original quantities separately and summing the resulting expressions. It is this principle which allows us to separate common factors from the terms of a polynomial.

Just as with numbers, an algebraic expression is a prime factor if it has no other factors except itself and 1. The factor $x^2 - 4$ is not prime, since it can be separated into $x - 2$ and $x + 2$.

The factors $x - 2$ and $x + 2$ are both prime factors, rime they cannot be separated into other factors.

The process of finding the factors of a polynomial is called FACTORING. An expression is said to be factored completely when it has been separated into its prime factors. The polynomial $3x^3 - 12x$ is factored completely as follows:

$$3x^3 - 12x = 3x(x - 2)(x + 2)$$

COMMON FACTORS

Factoring any polynomial begins with the removal of common factors. Notice that "removal" of a factor does not mean discarding it. To remove a factor is to insert parentheses and move the factor outside the parentheses as a common multiplier. The removal of common factors proceeds as follows:

1. Inspect the polynomial and find the factors which are common to all terms. These common factors, multiplied together, comprise the 'largest common factor."
2. Mentally divide each term of the polynomial by the largest common factor and write the quotients within a set of parentheses.
3. Write the largest common factor outside the parentheses as a common multiplier.

For example, the expression $x^2y - xy^2$ contains xy as a factor of each term. Therefore, it is factored as follows:

$$x^2y - xy^2 = xy(x - y)$$

Other examples of factoring by the removal of common factors are found in the following expressions:

$$6m^4n + 3m^3n^2 - 3m^2n^3 = 3m^2n(2m^2 + mn - n^2$$

$$-5x^2 - 15z = -5z(z(z + 3)$$

$$7x - 7y + 7z = 7(x - y + z)$$

In selecting common factors, always remove as many factors as possible from each term in order to factor completely. For example, x is a factor of $3ax^2 - 3ax$, so that $3ax^2 - 3ax$ is equal to $x(3ax - 3a)$. However, 3 and a are also factors. Thus the largest common factor is $3ax$. When factored completely, the expression is as follows:

$$3ax^2 - 3ax = 3ax(x - 1)$$

Practice problems: Remove the common factors:

1. $y^2 - y$
2. $a^3b^2 - a^2b^2$
3. $2b^3 - 8b^2 - 6b$
4. $6mn^2 + 30m^2n$
5. $\frac{2}{3}x - \frac{1}{3}y + \frac{1}{3}$

Answers:

1. $y(y - 1)$
2. $a^2b^2 - (a - 1)$
3. $2b\,(b^2 - 4b - 3)$
4. $6mn\,(n + 30\,m)$
5. $\frac{2}{3}x - \frac{1}{3}y + \frac{1}{3}$

LITERAL EXPONENTS

It is frequently necessary to remove common factors involving literal exponents; that is, exponents composed of letters rather than numbers. A typical expression involving literal exponents is $x^{2a} + x^{a}$,' in which x^{a} is a common factor. The factored form is $x^{a}(x^{a} + 1)$. Another example of this type is a^{m+n}, $+ 2a^{m}$. Remember that $(a^{m \cdot n})$ is equivalent to $(a^{m})(a^{n})$. Thus the factored form is as follows:

$$a^{m+n} + 2a^{m} = a^{m}.\ a^{n} + 2a^{n} = a^{m}(a^{n} + 2)$$

BINOMIAL FORM

The distinctions between monomial, binomial, and trinomial factors are discussed in detail in chapter 9 of this course. An expression such as $a(x + y) + b(x + y)$ has a common factor in binomial form. The factor $(x + y)$ can be removed from both terms, with the following result:

$$a(x + y) + b(x + y) - (x + y)(a + b)$$

Sometimes it 16 easier to see this if a single letter is substituted temporarily for the binomial, Thus, let $(x + y) - n$, so that $a(x + y) + b(x + y)$ reducer to $(an + bn)$. The factored form is $n(a + b)$, which becomes $(x + y)(a + b)$ when n is replaced by its equal, $(x + y)$.

Another form of this type is $x(y - 2) - w(a - y)$. Notice that this expression could be factored easily if the binomial in the Second term were $(y - z)$. We can show that $- w(z - y)$ is equalivalent to $w(z - y)$, as follows:

$$\begin{aligned} -w(z - y) &= -w[(-1).(-1).\ z + (-1),\ y] \\ &= -w[(-1).(-1).\ z + y]\} \\ &= -w(-1)[-z + y] = -w(y - z) \end{aligned}$$

Substituting $+w(y - z)$ for $-w(z - y)$ in the original expression, we may now factor as follows:

$$x(y - z) = w(z - y) = x(y - z) + w(y - z) = x(y - z)\ (x - w)$$

In factoring an expression such as $ax + bx + ay + by$, common monomial factors are removed first, as follows:

$$ax - bx + ay + by = x(a + b) + y(a + b)$$

Having removed the common monomial factors, we then remove the common binomial factor to obtain $(a + b)(x + y)$. Notice that we could have rewritten the expression as $ax + ay + bx + by$, based on the commutative law of addition, which states that the sum of two or more terms is the same regardless of the order in which they are arranged. The first step in factoring would then produce $a(x + y) + b(x + y)$ and the final form would be $(x + y)(a + b)$. This is equivalent to $(a + b)(x + y)$, by the commutative law of multiplication, which states that the product of two or more factors is the same regardless of the order in which they are arranged.

Practice problems: Factor each of the following:

1. $x^{3a} + 3x^{2a}$

2. $xy^2 + y + x^2y + x$
3. $e^x + 4e^{4x}$
4. $7(x^2 + y^2) - 3z(x^2 + y^2)$
5. $a^2 + ab - ac - cb$
6. $\frac{1}{2}e^2r - \frac{1}{6}er^2$

Answers:

1. $x^{2a}(x^a + 3)$
2. $(xy + 1)(x + y)$
3. $e^x(1 + 4e^{3x})$
4. $(x^2 + y^2)(7 + 3z)$
5. $(a + b)(a - b)$
6. $\frac{1}{2}er(e - \frac{1}{3}r)$

INOMIAL FACTORS

After any common factor has been removed from a polynomial, the remaining polynomial factor must be examined further for other factors. Skill in factoring is principally the ability to recognize certain types of products such as the square of a sum or difference. Therefore, it is important to be familiar with the special products discussed in chapter 9.

DIFFERENCE OF TWO SQUARES

In chapter 9 we learned that the product of the sum and difference of two numbers is the difference of their squares. Thus, $(a + b)(a - b) = a^2 - b^2$. Conversely, if a binomial is the difference of two squares, its factors are the sum and difference of the square roots. For example, in $9a^2 - 4b^2$ both $9a^2$ and $4b^2$ are perfect squares. The square roots are $3a$ and $2b$, respectively. Connect these square roots with a plus sign to get one factor of $9a^2 - 4b^2$ and with a minus sign to get the other factor. The two binomial factors are $3a - 2b$ and $3a + 2b$. Therefore, factored completely, the binomial can be written as follows:

$$9a^2 - 4b^2 = (3a - 2b)(3a + 2b)$$

We may check to see if these factors are correct by multi $\frac{1}{2}er(e - \frac{1}{3}r)$ to see if their product is the original binomial. The expression $20xy^3 - 5xy^3$ reduces to the difference of two squares after the common factor $5xy$ is removed. Completely factored, this expression produces the following:

$$20x^3y - 5xy^3 = 5xy(4x^2 - y^2)$$

$$5xy(2x - y)(2x + y)$$

Other examples that show the difference of two squares in factored form are as follows:

$$49 - 16 = (7 + 4)(7 - 4)$$

$$16a^2 - 4x^2 = 4(4a^2 - x^2) = 4(2a + x)\ (2a - x)$$

$$4x^2y = 9y = y(4x^2 - 9)$$

$$y(2x + 3)\ (2x - 3)$$

Practice problems: Factor each of the following:

1. $a^2 - b^2$
2. $a^2 - b^2$
3. $a^2\ b^2 - 1$
4. $a^2 - 144$
5. $x^2 - y^2$
6. $y^2 - 36$
7. $1 - 4y^2$
8. $9a^2 - 16$

Answers:

1. $(a + b)(a - b)$
2. $(a + 3)(a - 3)$
3. $(ab + 3)(ab - 1)$
4. $(a + 12)(a - 12)$
5. $(x + y)(x - y)$
6. $(y + 6)(y - 6)$

7. $(1-2y)(1-2y)$
8. $(3a+4)(3a-4)$

SPECIAL BINOMIAL FORMS

Special cases involving binomial expressions are frequently encountered. All such expressions may be factored by reference to general formulas, but these formulas are beyond the scope of this course. For our purposes, analysis of some special cases will be sufficient.

Even Exponents

When the exponents on both terms of the binomial are even, the expression may be treated as the sums or difference of two squares. For example, $x^6 - y^6$ can be rewritten as $(x^3)^2 - (y^3)^2$ which results in the following factored..

$$(x^6 - y^6) = (x^3 - y^3)(x^3 + y^3)$$

In general, a binomial with even exponents has the form $x^{2m} \pm y^{2n}$, since all even numbers have 2 as a factor. If the connecting sign is positive, the expression may not be factorable; for example, $x^2 + y^2$, $x^4 + y^4$, and $x^6 + y^6$ are all nonfactorable binomials. If the.connecting sign is negative, a binomial with even exponents is factorable as follows:

$$x^{2m} - x^{2n} = (x^m - y^n)(x^m + y^n)$$

A special case which is particularly important because it occurs so often is the binomial which has the numeral 1 as one of its terms. For example, $x^4 - 1$ is factorable as the difference of two squares, as follows:

$$x^4 - 1 = (x^2 - 1)(x^2 + 1) = (x - 1)(x + 1)(x^2 + 1)$$

Odd Exponents

Two special cases involving odd exponents are of particular importance. These are the sum of two cubes and the difference of two cubes. Examples of the sum and difference of two cubes, showing their factored forms, are as follows:

$$x^3 + y^3 = (x + y)(x^2 - xy + y^2)$$

$$x^3 - y^3 = (x - y)(x^2 + xy + y^2)$$

Notice that each of these factored forms involves a first degree binomial factor ($(x + y)$ in the first case and $(x - y)$ in the second). The connecting sign in the first degree binomial factor corresponds to the connecting sign in the. original unfactored binomial.

We are now in a position to give the completely factored form of $x^6 - y^6$, as follows:

$$x^6 + y^6 = (x^3 - y^3)(x^3 - y^3) = (x - y)(x^2 + xy + y^2)$$
$$(x + y)(x^2 - xy + y^2)$$

In general, $(x + y)$ is a factor of $(x^n + y^n)$ if n is odd. If n is even, $(x^n + y^n)$ is not factorable unless it can be expressed as the sum of two cubes. When the connecting sign is negative, the binomial is always factorable if n is a whole number greater than 1.

That is, $(x - y)$ is a factor of $(x^n - y^n)$ for both odd and even values of n. The special case in which one of the terms of the binomial is the numeral 1 occurs frequently. An example of this is $x^3 + 1$, which is factorable as the sum of two cubes, as follows:

$$x^3 + 1 = (x + 1)(x^2 - x + 1)$$

In a similar manner, 1 + x6 can be treated as the sum of two cubes and factored as follows:

$$1 + x^6 -1 + (x^2)^3 = (1+ x^2)(1-x^2 + x^4)$$

Practice problems: In each of the following problems, factor completely:

1. $x^4 - y^4$
2. $m^3 - n^3$
3. $x^6 - y^6$
4. $x^3 - y^3$
5. $a^9 - b^9$
6. $x^{2a} - y^{2b}$
7. $1 - x^4$

8. $x^6 + 1$
9. $1 - x^3$

Answers:

1. $(x + y)(x - y)(x^2 + y^2)$
2. $(m + n)(m^2 - mn + n^2)$
3. $(x + y)(x - y)(x^2 + xy^2 + y^2)(x^2 - xy + y^2)$
4. $(x - y)\ (x^2 + xy + y^2)$
5. $(a - b)(a^2 + ab + b^2)(a^6 + a^3b^3 + b^6)$
6. $(x^a - y^b)(x^a + y^b)$
7. $(1 + x^2)(1 - x)(1 + x)$
8. $(x^2 + 1)(x^4 - x^2 + 1)$
9. $(1 - x)(1 + x + x^2)$

TRINOMIAL SQUARES

A trinomial that is the square of a binomial is called a Trinomial Square. Trinomials that are perfect squares factor into either the square of a sum or the square of a difference. Recalling that $(x + y)^2 = x^2 + 2xy + y^2$ and $(x - y)^2 = x^2 - 2xy + y^2$, the form of a trinomial square is apparent. The first term and the last term are perfect squares and their signs are positive. The middle term is twice the product of the square roots of these two numbers. The sign of the middle term is plus if a sum has been squared; it is minus if a difference has been squared.

The polynomial $16x^2 - 8xy + x^2$ is a trinomial in which the first term, $16x$, and the last term, y^2, are perfect squares with positive signs. The square roots are $4x$ and y. Twice the product of these square roots is $2(4x)(y) = 8xy$. The middle term is preceded by a minus sign indicating that a difference has been squared. In factored form this trinomial is as follows:

$$16x^2 - 8xy + y^2 = (4x - y)^2$$

To factor the trinomial, we simply take the square roots of the end terms and join them with a plus sign if the middle term is preceded by a plus or with a minus if the middle term is preceded by a minus.

The terms of a trinomial may appear in any order. Thus, $8xy + y^2 + 16x^2$ is a trinomial square and may be factored as follows:

$$8xy + y^2 + 16x^2 = 16x^2 + 8xy + y^2 = (4x + y)^2$$

Practice problems: Among the following expressions, factor those which are trinomial squares:

1. $y^2 - 8y + 16$
2. $16y^2 + 30x + 9$
3. $36 + 12x + x^2$
4. $a^2 + 2ab + b^2$
5. $12y + 9y^2 + 4xy$6. $36 + 12x + x^2$
7. $9 - 6cd + c^2d^2$ 8. $x4 + 4x^2 + 4$

Answers:

1. $(y - 4)^2$
2. Not a trinomial square
3. $(6 + x)^2$
4. $(a + b)^2$
5. Not a trinomial square
6. $(3 - cd)^2$
7. $(3 - cd)^2$
8. $(x^2 + 2)^2$

SUPPLYING THE MISSING TERM

Skill in recognizing trinomial squares may be improved by practicing the solution of problems which require supplying a missing term. For example, the expression $y^2 + (?) + 16$ can be made to form a perfect trinomial square by supplying the correct term to fill the parentheses.

The middle term must be twice the product of the square roots of the two perfect square terms; that is, $(2)(4)(y)$, or $8y$. Check: $y^2 + 8y + 16 = (y + 4)^2$. The missing term is $8y$.

Suppose that we wish to supply the missing term in $16x^2 +$

24xy + (?) so that the three terms will form a perfect trinomial square. The square root of the first term is $4x$. One-half the middle term is $12xy$. Divide $12xy$ by $4x$. The result is $3y$ which is the square root of the last term. Thus, our missing term is $9y^2$. Checking, we find that $(4x + 3y)^2 = 16x^2 + 24xy + 9y^2$.

Practice problems: In each of the following problems, supply the missing term to form a perfect trinomial square:

1. $x^2 + (?) + y^2$
2. $t^2 + (?) + y^2$
3. $9a^2 - (?) + 25b^2$
4. $4m^2 + 16m + (?)$
5. $x^2 + 4x + (?)$
6. $c^2 - 6cd + (?)$

Answers:

1. $2xy$
2. $10t$
3. $30ab$
4. 16
5. 4
6. $9d^2$

OTHER TRINOMIALS

It is sometimes possible to factor trinomials that are not perfect squares. Following are some examples of such trinomials, and the expressions of which they are products:

1. $(x + 3)(x + 4) = x^2 + 7x + 12$
2. $(x - 3)(x - 4) = x^2 - 7x + 12$
3. $(x - 3)(x + 4) = x^2 + 7x - 12$
4. $(x + 3)(x - 4) = x^2 - 7x - 12$

It is apparent that trinomials like these may be factored into binomials as shown. Notice how the trinomial in each of the preceding examples is formed. The first term is the square of the common term of the binomial factors. The second term is the algebraic sum of their unlike terms times their common term.

The third term is the product of their unlike terms, such trinomials may be factored as the product of two binomials if there are two numbers such that their algebraic sum is the coefficient of the middle term and their product is the last term.

For example, let us factor the expression $x^2 - 12x + 32$. If the expression is factorable, there will be a common term, x, in each of the binomial factors. We begin factoring by placing this term within each set of parentheses, as follows:

$$(x)(x)$$

Next, we must find the other terms that are to go in the parentheses. They will be two numbers such that their algebraic sum is –12 and satisfy the conditions. Thus, the following expression results:

$$x^2 - 12x + 32 = (x - 8)(x - 4)$$

It is of value in factoring to note some useful facts about trinomials. If both the second and third terms of the trinomial are positive, the signs of the terms to be found are positive as in example 1 of this section.

If the second term is negative and the last is positive, both terms to be found will be negative as in example 2. If the third term of the trinomial is negative, one of the terms to be found is positive and the other is negative as in examples 3 and 4. Concerning this last case, if the second term is positive as in example 3, the positive term in the factors has the greater numerical value. If the second term is negative as in example 4, the negative term in the factors has the greater numerical value.

It should be remembered that not all trinomials are factorable. For example, $x^2 + 4x + 2$ cannot be factored since there are no two rational numbers whose product is 2 and whose sum is 4.

Practice problems: Factor completely, in the following problems:

1. $y^2 + 15y + 50$
2. $y^2 + 2y + 24$
3. $x^2 + 8x - 48$
4. $x^2 - 4x - 60$
5. $x^2 + 12x - 45$
6. $x^2 - 15x + 56$
7. $x^2 + 2x - 48$
8. $x^2 + 14x + 24$

Answers:

1. $(y + 5)(y + 10)$
2. $(y + 6)(y + 4)$
3. $(x - 12)(x - 4)$
4. $(x - 10)(x + 6)$
5. $(x - 15)(x + 3)$
6. $(x - 7)(x - 8)$
7. $(x - 6)(x + 8)$
8. $(x + 12)(x + 2)$

Thus far we have considered only those expressions in which the coefficient of the first term is 1. When the coefficient of the first term is other than 1, the expression can be factored as shown in the following example:

$$6x^2 - x - 2 = (2x + 1)(3x - 2)$$

Although this result can be obtained by the trial and error method, the following procedure saves time and effort.

First, find two numbers whose sum is the coefficient of the second term (–1 in this example) and whose product is equal to the product of the third term and the coefficient of the first term (in this example, (6)(–2) or –12). By inspection, the desired numbers are found to be –4 and +3.

Using these two numbers as coefficients for x, we can rewrite the original expression as $6x^2 - 4x + 3x - 2$ and factor as follows:

$6x^2 - 4x + 3x - 2 = 2x(3x - 2) + 1(3x - 1) = (2x + 1)(3x - 2)$

Practice problems: Factor completely, in the following problems:

1. $2x^2 + 13x + 21$ 2. $16x^2 + 26x + 3$

3. $15x^2 - 8x - 15$ 4. $12x^2 - 8x - 15$

Answers:

1. $(2x + 7)(x + 3)$
2. $(2x + 3)(8x + 1)$
3. $(3x + 1)(5x - 7)$
4. $(6x + 5)(2x - 3)$

REDUCING FRACTIONS TO LOWEST TERMS

There are many useful applications of factoring. One of the most important is that of simplifying algebraic fractions. Fractions that contain algebraic expressions in the numerator or denominator, or both, can be reduced to lower terms, if there are factors common to numerator and denominator. If the terms of a fraction are monomials, common factors are immediately apparent, as in the following expression:

$$\frac{3x^2y}{6xy} = \frac{3xy(x)}{3xy(2)} = \frac{x}{2}$$

If the terms of a fraction are polynomials, the polynomials must be factored in order to recognize the existence of common factors, as in the following two examples:

1. $$\frac{a-b}{a^2 - 2ab + b^2} = \frac{a-b}{(a-b)(a-b)} - \frac{1}{(a-b)}$$

2. $$\frac{4x^2 - 9}{6x^2 - 9x} = \frac{(2x+3)(2x-3)}{3x(2x-3)} = \frac{(2x+3)}{3x}$$

Notice that without the valuable process of factoring, we would be forced to use the fractions in their more complicated form. When there are factors common to both numerator and denominator, it is obviously more practical to cancel them (first using the factoring process) before proceeding.

Practice problems: Reduce to lowest terms in each of the following:

1. $\dfrac{12}{6x+12}$

2. $\dfrac{a^2-b^2}{a^2-2ab+b^2}$

3. $\dfrac{y^2-14y+45}{y^2-8y-9}$

4. $\dfrac{y^2-25}{y^2-8y+15}$

5. $\dfrac{a^2-5a-24}{a^2-64}$

6. $\dfrac{4x^2y-9y}{4x^2+12x+9}$

Answers:

1. $\dfrac{2}{x+2}$

2. $\dfrac{a+b}{a-b}$

3. $\dfrac{y-5}{y+1}$

4. $\dfrac{y+5}{y-3}$

5. $\dfrac{a+3}{a+8}$

6. $\dfrac{y(2x-3)}{2x+3}$

OPERATIONS INVOLVING FRACTIONS

Addition, subtraction, multiplication, and division operations involving algebraic fractions are often simplified by means of factoring, whereas they would be quite complicated without the use of factoring.

MULTIPLYING FRACTIONS

Multiplication of fractions that contain polynomials is similar to multiplication of fractions that contain only arithmetic numbers. If this fact is kept in mind, the student will have little difficulty in mastering multiplication in algebra. For instance, we recall that to multiply a fraction by a whole number, we simply multiply the numerator by the whole number. This,is illustrated in the following example:

Arithmetic: $4\times\frac{3}{17}-\frac{12}{17}$

Algebra: $(x-4)-\frac{3}{x^2-5}=\frac{3x-12}{x^2-5}$

Sometimes the work may be simplified by factoring and canceling before carrying out this multiplication. The following example illustrates this:

$$(2a-8).\frac{3}{a^2-8a+18}=\frac{2(a-4)}{1}.\frac{3}{(a-4)(a-4)}=\frac{2(3)}{a-4}=\frac{6}{a-4}$$

When the multiplier is a fraction, the ruler of arithmetic remain applicable-that is, multiply numerators together and denominators together, This is illustrated,as follows:

Arithmetic: $\frac{4}{5}\times\frac{2}{3}=\frac{8}{15}$

Algebra: $\frac{a+b}{a-b}.\frac{a}{a-b}=\frac{a(a+b)}{(a-b)}$

Where possible, the work may be considerably reduced by factoring, canceling, and then carrying out the multiplication, as in the following example:

$$\frac{x^2-2x+1}{x^2-9}\cdot\frac{x^2+x-6}{x^2-1}=\frac{(x-1)(x-1)}{(x-3)(x-3)}\cdot\frac{(x+3)(x-2)}{(x+1)(x-1)}$$

$$=\frac{(x-1)(x-3)}{(x-3)(x+1)}=\frac{x^2-3x+2}{x^2-2x-3}$$

Although the factors may be multiplied to form two trinomials as shown, it is usually sufficient to leave the answer in factored form. Practice problems. In the following problems, multiply as indicated:

1. $5a^2.\frac{3b}{a+b}$
2. $\frac{x+y}{x^2}\cdot\frac{x-y}{x-1}$
3. $\frac{a^2+2ab+b^2}{a^2-b^2},\frac{6a}{3a+3b}$
4. $\frac{a-1}{2a^2+4a+2}\cdot\frac{(a+1)^2}{a-1}$

Answers:

1. $\frac{15a^2b}{a+b}$
2. $\frac{x^2-y^2}{x^3-x^2}$
3. $\frac{2a}{(a-b)}$
4. $\frac{1}{2}$

DIVIDING FRACTIONS

The ruler of arithmetic apply to the division of algebraic fractions; as in arithmetic, simply invert the divisor and multiply, as follows:

Arithmetic:

$$\frac{3}{8}+\frac{9}{16}=\frac{3}{8}\times\frac{16}{9}=\frac{3}{8}\times\frac{(8)(2)}{(3)(3)}=\frac{2}{3}$$

Algebra:

$$\frac{x-3y}{x+3y}+\frac{x^2-6xy+9y^2}{x^2+7xy+12y^2}$$

$$\frac{x-3y}{x+3y}+\frac{x^2+7xy+12y^2}{x^2-6xy+9y^2}$$

$$=\frac{x-3y}{x-3y}\cdot\frac{(x+3y)(x+4y)}{(x-3y)(x-3y)}=\frac{x+4y}{x-3y}$$

Practice problems. In the following problems, divide and reduce to lowest terms:

1. $\dfrac{x-2}{x^2+4x+4}\div\dfrac{1}{x^2-4}$
2. $\dfrac{3a-1}{a^3+3a}\div\dfrac{a+1}{a^2+3}$
3. $\dfrac{a^3-4a^2+3a}{a+2}\div(a-3)$
4. $\dfrac{6t+12}{9t^2+6t-24}\div\dfrac{8t-12}{15t-20}$

Answers:

1. $\dfrac{(x-2)^2}{x+2}$
2. $\dfrac{2a-1}{a^2+a}$
3. $\dfrac{a(a-1)}{a+2}$
4. $\dfrac{5}{4t-6}$

ADDING AND SUBTRACTING FRACTIONS

The rules of arithmetic for adding and subtracting fractions are applicable to algebraic fractions. Fractions that are to be combined by addition or subtraction must have the same denominator, The numerators are then combined according to the operation indicated and the result is placed over the denominator.

For example, in the expression

$$\frac{x-4}{x-2}+\frac{2-11x}{2-x}$$

the second denominator will be the same as the first, if its sign is changed, The value of the fraction will remain the same if the sign of the numerator is also changed, Thus, we have the following simplification:

$$\frac{x-4}{x-2}+\frac{2-11x}{2-x}=\frac{x-4}{x-2}+\frac{-(2-11x)}{-(2-x)}$$

$$=\frac{x-4}{x-2}+\frac{11(x-2)}{x-2}=\frac{x-4+11x-2}{x-2}$$

$$=\frac{12x-6}{x-6}=\frac{6(2x-1)}{x-2}$$

When the denominators are not the same, we must reduce all fractions to be added or subtracted to a common denominator and then proceed.

Consider, for example,

$$\frac{4}{x^2-4}+\frac{3}{x^2-4x-12}$$

We first must find the least common denominator (LCD). Remember this is the least number that is exactly divisible by each of the denominators.

To find such a number, as in arithmetic, we first separate each of the denominators into prime factors. The LCD will contain all of the various prime factors, each one as many times as it occurs in any of the denominators. Factoring, we have

$$\frac{4}{(x+2)(x-2)}+\frac{3}{(x-6)(x+2)}$$

and the LCD is $(x+2)(x-2)(x-6)$. Rewriting the fractions with this denominator and adding numerators, we have the following expression:

$$\frac{4(x-6)}{(x+2)(x-2)(x-6)}+\frac{3(x-2)}{(x+2)(x-2)(x-6)}$$

$$=\frac{4(x-6)+3(x-2)}{LCD}$$

$$=\frac{4x-24+3x-6)}{LCD}$$

$$=\frac{7x-30}{(x+2)(x-2)(x-6)}$$

As another example, consider. Factoring the denominator of the second fraction, we find that the LCD is (x + 3)(x + 1). Rewriting the original fractions with the LCD as denominator, we may now combine the fractions as follows:

$$\frac{4(x+1)}{(x+3)(x+1)}-\frac{(x+2)}{(x+3)(x+1)}$$

$$=\frac{4x+4-x-2}{(x+3)(x+1)}=\frac{3x+2}{(x+3)(x+1)}$$

Practice problems: Perform the indicated operations in each of the following problems:

1. $\frac{3x-4}{x^2+x-2}-\frac{x-2}{x-1}$

2. $\frac{3a}{a^2-9}-\frac{3}{3-a}$

3. $\frac{x-3}{3x}+\frac{x+2}{2x}$

4. $\frac{1}{a^4-1}=\frac{1}{a+1}$

5. $\dfrac{3}{(a+4)^2} - \dfrac{2}{a(a+4)} + \dfrac{1}{6(a+4)}$

Answers:

1. $\dfrac{3x - x^2}{(x+2)(x-1)}$

2. $\dfrac{6a+9}{(a+3)(a-3)}$

3. $\dfrac{5}{6}$

4. $\dfrac{2 - a^3 + a^2 - a}{(a^2+1)(a+1)(a-1)}$

5. $\dfrac{a^2 + 10a - 48}{6a(a+4)^2}$

Chapter 7

Linear Equations

LINEAR EQUATIONS ONE VARIABLE

One of the principal reasons for an intensive study of polynomials, grouping symbols, factoring, and fractions is to prepare for solving equations. The equation is perhaps the most important tool in algebra, and the more skillful the student becomes in working with equations, the greater will be his ease in solving problems.

Before learning to solve equations, it is necessary to become familiar with the words used in the discussion of them. An Equation is a statement that two expressions are equal in value. Thus,

$$4 + 5 = 9$$

and

$$A = lw$$

(Area of a rectangle = length × width),

are equations. The part to the left of the equality sign is called the Left Member, or first member, of the equation. The part to the right is the Right Member, or second member, of the equation.

The members of an equation are sometimes thought of as corresponding to two weights that balance a scale. This compari8on is often helpful to students who are learning to solve equations. It is obvious, in

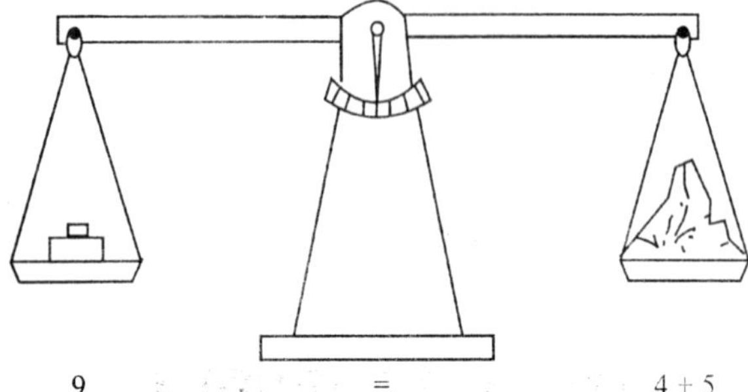

9 = 4 + 5

Fig. Equation Compared to a Balance Scale

The case of the scale, that any change made in one pan must be accompanied by an equal change in the other pan. Otherwise the scale will not balance. Operations on equations are based on the same principle. The members must be kept balanced or the equality is lost.

CONSTANTS AND VARIABLES

Expressions in algebra consist of constants and variables. A Constant is a quantity whose value remains the same throughout a particular problem. A Variable is a quantity whose value is free to vary. There are two kinds of constants-fixed and arbitrary. Numbers such as 7, – 3, 1/2, and p are examples of Fixed constants. Their values never change. In $5x + 7 = 0$, the numbers 0, 5, and 7, are fixed constants.

Arbitrary constants can be assigned different values for different problems. Arbitrary constants are indicated by letters- quite often letters at the beginning of the alphabet such as *a, b, c,* and *d*. In

$$ax + b = 0,.$$

the letters *a* and *b* represent arbitrary constants. The form *ax t b* – 0 represent many linear equations. If we give *a* and *b* particular values, say *a* – 5 and *b* = 7, then these constants become fixed, for this particular problem, and the equation becomes

$$5x\ t\ 7 = 0$$

A variable may have one value or it may have many values in a dlscuseion. The letters at the end of the alphabet, such as *x*,

y, z, and w, usually are used to represent variables. In $5x + 7$, the letter x is the variable. If $x = 1$, then

$$5x + 7 = 5 + 7 = 12$$

If $x = 2$, then

$$5x + 7 = 5(2) + 7 = 10 + 7 = 17$$

and so on for as many values of x as we desire to select.

If the expression $5x + 7$ is set equal to some particular number, say –23, then the resulting equality

$$5x + 7 = -23$$

holds true for just one value of x. The value is –6, since

$$5(-6) + 7 = -23$$

In an algebraic expression, terms that contain a variable are called Variable Terms. Terms that do not contain a variable are Constant Terms. The expression $5x + 7$ contains one variable term and one constant term. The variable term is $5x$, while 7 is the constant term. In $ax + b$, ax is the variable term and b is the constant term. A variable term often is designated by naming the variable it contains. In $5x + 7$, $5x$ is the x-term. In $ax + by$, ax is the x-term, while by is the y–term.

DEGREE OF AN EQUATION

The degree of an equation that has not more than one variable in each term is the exponent of the highest power to which that variable is raised in the equation. The equation

$$3x - 17 = 0$$

is a First-Degree equation, since x is raised only to the first power.

An example of a Second -Degree equation is

$$5x^2 - 2x + 1 = 0.$$

The equation,

$$4x^3 - 7x^2 = 0,$$

is of the Third Degree.

The equation,

$$3x - 2y = 5$$

is of the first degree in two variables, x and y. When more than one variable appears in a term, as in $xy = 5$, it is necessary to add the exponents of the variables within a term to get the degree of the equation. Since $1 + 1 = 2$, the equation $xy = 5$ is of the second degree.

LINEAR EQUATIONS

Graphs are used in many different forms to give visual pictures of certain related facts. For example, they are used to show business trends, production output, continued individual attainment, and so forth.

We find bar graphs, line graphs, circle graphs, and many other types, each of which is used for a particular need. In algebra, graphs are also used to give a visual picture containing a great deal of information about equations. Sometimes many numerical values, when substituted for the variables of an equation, will satisfy the conditions of the equation. On a particular type of graph several of these values are plotted (located), and when enough are plotted, a line is drawn through these points. For each particular equation a certain type of curve results. For equations in the first degree in one or two variables, the resulting shape of the "curve" is a straight line..

Thus, the name Liner Equation is derived. Equations of a higher degree form various other shapes. The name "linear equation" now applies to equations of the first degree, regardless of the number of variables they contain.

IDENTITIES

If a statement of equality involves one or more variables, it may be either an Identity (identical equation) or a Conditional Equation. An identity is an equality that states a fact, such as the following examples:

1. $9 + 5 = 14$

2. $2n + 5n = 7n$
3. $6(x - 3) = 6x - 18$

Notice that equation 3 merely shows the factored form of $6x - 18$ and holds true when any value of x is substituted. For example, if $x = 5$, it becomes

$$6(5 - 3) = 6(5) - 18$$
$$6(2) = 30 - 18$$
$$12 = 12$$

If x assumes the negative value – 10, this identity becomes

$$6(-10 - 3) = 6(-10) - 18$$
$$6(-13) = -60 - 18$$
$$-78 = -78$$

An identity is established when both sides of the equality have been reduced to the same number or the same expression. When 5 is substituted for x, the value of either side of $6(x - 3) = 6x - 18$ is 12. When –10 is substituted for x, the value on either side is –78. The fact that this equality is an identity can be shown also by factoring the right side so that the equality becomes

$$6(x - 3) = 6(x - 3)$$

The expressions on the two sides of the equality are identical.

CONDITIONAL EQUATIONS

A statement such as $2x - 1 = 0$ is an equality only when x has one particular value. Such a statement is called a Conditional Equation, since it is true only under the condition that $x = 1/2$. Likewise, the equation $y - 7 = 8$ holds true only if $y = 15$.

The value of the variable for which an equation in one variable holds true is a Root, or Solution, of the equation. When we speak of solving equations in algebra, we refer to conditional equations. The solution of a conditional equation can be verified by substituting for the variable its value, as determined by the solution.

The solution. is correct if the equality reduces to an identity. For example, if 1/2 is substituted for x in $2x - 1 = 0$, the result is

$$2\left(\frac{1}{2}\right) - 1 = 0$$

$$1 - 1 = 0$$

$$0 = 0 \text{ (an identity)}$$

The identity is established for $x = 1/2$, since the value of each side of the equality reduces to zero

SOLVING LINEAR EQUATIONS

Solving a linear equation in one variable means finding the value of the variable that makes the equation true. For example, 11 is the Solution of $x - 7 = 4$, since $11 - 7 = 4$. The number 11 is said to Satisfy the equation. Basically, the operation used in solving equations is to manipulate both members, by addition, subtraction, multiplication, or division until the value of the variable becomes apparent.

This manipulation may be accomplished in a straightforward manner by use of the axioms outlined in chapter 3 of this course. These axioms may be summed up in the following rule: If both members of an equation are increased, decreased, multiplied, or divided by the same number, or by equal numbers, the results will be equal. (Division by zero is excluded.)

As mentioned earlier, an equation may be compared to a balance. What is done to one member must also be done to the other to maintain a balance. An equation must always be kept in balance or the equality is lost.

We use the above rule to remove or adjust terms and coefficients until the value of the variable is discovered. Some examples of equations solved by means of the four operations mentioned in the rule are given in the following paragraphs.

ADDITION

Find the value of x in the equation

$$x - 3 = 12$$

As in any equation, we must isolate the variable on either the right or left side. In this problem, we leave the variable on the left and perform the following steps:

1. Add 3 to both members of the equation, as follows:

$$x - 3 + 3 = 12 + 3$$

In effect, we are "undoing" the subtraction indicated by the expression $x - 3$, for the purpose of isolating x in the left member.

2. Combining terms, we have

$$x = 15$$

SUBTRACTION

Find the value of x in the equation

$$x + 14 = 24$$

1. Subtract 14 from each member. In effect, this undoes the addition indicated in the expression $x + 14$.

$$x + 14 - 14 = 24 - 14$$

2. Combining terms, we have

$$x = 10$$

MULTIPLICATION

Find the value of y in the equation

$$y/5 = 10$$

1. The only way to remove the 5 so that the y can be isolated is to undo the indicated division. Thus we use the inverse of division, which is multiplication. Multiplying both members by 5, we have the following:

$$5(y/5) = 5(10)$$

2. Performing the indicated multiplications, we have

$$y = 50$$

DIVISION

Find the value of x in the equation

$$3x = 15.$$

1. The multiplier 3 may be removed from the x by dividing the left member by 3. This must be balanced by dividing the right member by 3 also, as follows:
2. Performing the indicated divisions, we have

$$x = 5$$

Practice problems: Solve the following equations:

1. $m + 2 = 8$
2. $x - 5 = 11$
3. $6x = -48$
4. $\frac{x}{14} = 2$
5. $2n = 5$
6. $\frac{1}{6}y = 6$

Answers:

1. $m = 6$
2. $x = 16$
3. $x = -86$
4. $x = 28$
5. $n = 2\frac{1}{2}$
6. $y = 36$

SOLUTIONS REQUIRING MORE THAN ONE OPERATION

Most equations involve more steps in their solutions than the simple equations already described, but the basic operations remain unchanged. If the basic axioms are kept well in mind, these more complicated equations will not become too difficult.

Equations may require one or all of the basic operations before a solution can be obtained.

Subtraction and Division

Find the value of x in the following equation:

$$2x + 4 = 16$$

1. The term containing x is isolated on the left by subtracting 4 from the left member.

This operation must be balanced by also subtracting 4 from the right member, as follows:

$$2x + 4 - 4 = 16 - 4$$

2. Performing the indicated operations, we have

$$2x = 12$$

3. The multiplier 2 is removed from the x by dividing both sides of the equation by 2, as follows:

$$\frac{2x}{2} = \frac{12}{1}$$

$$x = 6$$

Addition, Multiplication, and Division

Find the value of y in the following equation:

$$\frac{3y}{2} - 4 = 11$$

1. Isolate the term containing y on the left by adding 4 to both sides, as follows:

$$\frac{3y}{2} - 4 + 4 = 11 + 4$$

$$\frac{3y}{2} = 15$$

2. Since the 2 will not divide the 3 exactly, multiply the left member by 2 in order to eliminate the fraction. This operation must be balanced by multiplying the right member by 2, as follows:

$$2\left(\frac{3y}{2}\right) = 2(15)$$

$$3y = 30$$

3. Divide both members by 3, in order to isolate the y in the left member, as follows:

$$\frac{3y}{3} = \frac{30}{3}$$

$$y = 10$$

Equations Having the Variable in More Than One Term

Find the value of x in the following equation:

$$\frac{3x}{4} + x = 12 - x$$

1. Rewrite the equation with no terms containing the variable in the right member. This requires adding x to the right member to eliminate the $-x$ term, and balance requires that we also add x to the left member, as follows:

$$\frac{3x}{4} + x + x = 12 - x + x$$

$$\frac{3x}{4} + 2x = 12$$

2. Since the 4 will not divide the 3 exactly, it is necessary to multiply the first term by 4 to eliminate the fraction. However, notice that this multiplication cannot be performed on the first term only; any multiplier which is introduced for simplification purposes must be applied to the entire equation. Thus each term in the equation is multiplied by 4, as follows:

$$4\left(\frac{3x}{4}\right) + 4(2x) = 4(12)$$

$$3x + 8x = 48$$

3. Add the terms containing x and then divide both

sides by 11 to isolate the x in the left member, as follows:

$$11x = 48$$

$$x = \frac{48}{11}$$

Practice problems: Solve each of the following equations:

1. $x - 1 = \frac{1}{2}$
2. $\frac{y}{3} + y = 8$
3. $\frac{x}{4} + 3x = 7$
4. $4 - 7x = 9 - 8x$
5. $\frac{y}{2} + 6y = 13$
6. $\frac{1}{2}x - 2x = 25 + x$

Answers:

1. $x - 3/2$
2. $y = 6$
3. $x = 28/13$
4. $x = 5$
5. $y = 2$
6. $x = -10$

EQUATIONS WITH LITERAL COEFFICIENTS

As stated earlier, the first letters of the alphabet usually represent known quantities (constants), and the last letters represent unknown quantities (variables). Thus, we usually solve for x, y, or 2. An equation such as,

$$ax - 8 = bx - 5$$

has letters as coefficients. Equations with literal coefficients are solved in the same way as equations with numerical coefficients, except that when an operation cannot actually be performed, it merely is indicated. In solving for x in the equation

$$ax - 8 = bx - 5$$

subtract bx from both members and add 8 to both members. The result is

$$ax - bx = 8 - 5$$

Since the subtraction on the left side cannot actually be performed, it is indicated. The quantity, $a - b$, is the coefficient of x when terms are collected. The equation takes the form

$$(a - b)\, x = 3$$

Now divide both sides of the equation by $a - b$. Again the result can be only indicated. The solution of the equation is

$$x = \frac{3}{a - b}$$

In solving for y in the equation

$$ay + b = 4$$

subtract b from both members as follows:

$$ay = 4 - b$$

Dividing both members by a, the solution is

$$y = \frac{4 - b}{a}$$

Practice problems: Solve for x in each of the following:

1. $3 + x = b$
2. $4x = 8 + t$
3. $3x + 6m = 7m$
4. $ax - 2(x + b) = 3a$

Answers:

1. $x = b - 3$
2. $x = \frac{8 + t}{4}$

3. $x = \frac{m}{3}$

4. $x = \frac{3a + 2b}{a - 2}$

REMOVING SIGNS OF GROUPING

If signs of grouping appear in an equation they should be removed in the manner indicated in this course. For example, solve the equation.

$$5 = 24 - [x - 12(x - 2) - 6(x - 2)]$$

Notice that the same expression, $x - 2$, occurs in both parentheses. By combining the terms containing $(x - 2)$, the equation becomes

$$5 = 24 - [x - 18(x - 2)]$$

Next, remove the parentheses and then the bracket, obtaining

$$\begin{aligned} 5 &= 24 - [x - 18x + 36] \\ &= 24 - [36 - 17x] \\ &= 24 - 36 + 17x \\ &= -12 + 17x \end{aligned}$$

Subtracting $17x$ from both members and then subtracting 5 from both members, we have

$$-17x = -12 - 5$$

$$-17x = -17$$

Divide both members by –17. The solution is

$$x = 1$$

EQUATIONS CONTAINING FRACTIONS

To solve for x in an equation such as first clear the equation of fraction, To do this, find the least common denominator of the fraction. Then multiply both rider of the equation by the LCD. The least common denominator of 3, 12, 4, and 2 is 12. Multiply both rider of the equation by 12. The resulting equation is

$$8x + x - 12 - 3 + 8x$$

Subtract $6x$ from both members, add 12 to both members, and collect like terms as follows:

$$9x - 6x = 12 + 3$$

$$3x = 15$$

The solution is

$$x = 5$$

To prove that $x = 5$ is the correct solution, substitute 5 for x in the original equation and show that both sides of the equation reduce to the same value. The result of substitution is

$$\frac{2(5)}{3} + \frac{5}{12} - 1 = \frac{1}{4} + \frac{5}{2}$$

In establishing an identity, the two sides of the equality are treated separately, and the op the equality, and it is desirable to find the least common denominator for more than one set of fractions. The same denominator could be used on both sides of the equality, but this might make some of the terms of the fractions larger than necessary.

Proceeding in establishing the identity for $x = 5$ in the foregoing equation we obtain

$$\frac{10}{3} + \frac{5}{12} - \frac{3}{3} = \frac{1}{4} + \frac{10}{4}$$

$$\frac{7}{3} + \frac{5}{12} = \frac{5}{12}$$

$$\frac{28}{12} + \frac{5}{12} = \frac{11}{4}$$

$$\frac{33}{12} = \frac{11}{4}$$

Each member of the equality has the value 11/4 when $x =$ 5. The fact that the equation be proves that $x = 5$ is the solution.

Practice problems: Solve each of the following equations:

1. $\frac{x}{4} - 2 = \frac{x}{6}$

2. $\frac{1}{2}-\frac{1}{v}=\frac{1}{3}$

3. $\frac{y}{2}-\frac{y}{3}=5$

4. $\frac{3}{4x}=6$

Answers:

1. $x = 24$
3. $y = 30$
2. $v = 6$
4. $x = 1/8$

GENERAL FORM OF A LINEAR EQUATION

The expression General Form, in mathematics, implies a form to which all expressions or equations of a certain type can be reduced. The only possible terms in a linear equation in one variable are the first-degree term and the constant term. Therefore, the general form of a linear equation in one variable is

$$ax + b = 0$$

By selecting various values for a and b, this form can represent any linear equation in one variable after such an equation has been simpli represents the numerical equation

$$7x + 5 = 0$$

If $a = 2m - n$ and $b = p - q$, then $ax + b = 0$ represents the literal equation

$$(2m - n)x + p - q = 0$$

This equation is solved as follows:

$$(2m - n)x + (p - q) - (p - q) = 0 - (p - q)$$

$$(2m - n)x = 0 - (p - q)$$

$$x = \frac{-(p-q)}{2m-n} = \frac{q-p}{2m-n}$$

USING EQUATIONS TO SOLVE PROBLEMS

To solve a problem, we first translate the numerical sense of the problem into an equation. To see how this is accomplished, consider the following examples and their solutions.

Example: Together Smith and Jones have $ 120. Jones has 5 times as much as Smith. How much has Smith?

Solution:

Step 1. Get the problem clearly in mind. There are two parts to each problem-what is given (the facts) and what we want to know (the question). In this problem. we know that Jones has 5 times as much as Smith and together they have $120. We want to know how much Smith has.

Step 2. Express the unknown as a letter. Usually we express the unknown or number we know the least about as a letter (conventionally we use x). Here we know the least about Smith's money. Let x represent the number of dollars Smith has.

Step 3. Express the other facts in terms of the unknown. If x is the number of dollars Smith has and Jones has 5 times as much, then $5x$ is the number of dollars Jones has.

Step 4. Express the facts as an equation. The problem will express or imply a relation between the expressions in steps 2 and 3. Smith's dollars plus Jones' dollars equal $120. Translating this statement into algebraic symbols, we have

$$x + 5x = 120$$

Solving the equation for x,

$$6x = 120$$

$$x = 20$$

Thus Smith has $20.

Step 5. Check: See if the solution satisfies the original statement of the problem. Smith and Jones have $120.

$$\underset{\text{(Smith's money)}}{\$20} + \underset{\text{(Jones' money)}}{\$100} = \&120$$

Example: Brown can do a piece of work in 5 hr. If Olsen can do it in 4 hr how long will it take them to do the work together?

Solution:

Step 1. Given: Brown could do 5 hr. Olsen could do it in 4 hours. the work in

Unknown: How long it takes them to do the work together.

Step 2. Let x represent the time it takes them to do the work together.

Step 3. Then $1/x$ is the amount they do together in 1 hr. Also, in 1 hour Brown does 1/5 of the work and Olsen does 1/4 of the work

Step 4. The amount done in 1 hr is equal to the part of the work done by Brown in 1 hr plus that done by Olsen in 1 hr.

$$\frac{1}{x}=\frac{1}{5}+\frac{1}{4}$$

Solving the equation,

$$20x\left(\frac{1}{x}\right)=20x\left(\frac{1}{2}\right)+20x\left(\frac{1}{4}\right)$$

$$20 = 4x + 5x \; 20 = 9x$$

$$\frac{20}{9}=x, \text{ or } x=2\frac{2}{9} \text{ hours}$$

They complete the work together in 2 2/9: hours.

Step 5. Check: $2\frac{2}{9}\times\frac{1}{5}$ = amount Brown does

$2\frac{2}{9}\times\frac{1}{5}$ = amount Olsen does

$$\left(\frac{20}{9}\times\frac{1}{5}\right)+\left(\frac{20}{9}\times\frac{1}{4}\right)=\frac{4}{9}+\frac{5}{9}=\frac{9}{9}$$

Practice problems: Use a linear equation in one variable to solve each of the following problems:

1. Find three numbers such that the second is twice the first and the third is three times as large as the first. Their sum is 180.
2. A seaman drew $ 75.00 pay in dollar bills and five-dollar bills. The number of dollar bills was three more than the number of five-draw? (Hint: If x is the number of five-dollar bills, then $5x$ is the number of dollars they represent.)
3. Airman A can complete a maintenance task in 4 hr. Airman B requires only 3 hr to do the same work. If they work together, how long should it take them to complete the job?

Answers:

1. First number is 30.

   Second number is 60.

   Third number is 90.
2. Number of five-dollar bills is 12.

   Number of one-dollar bills is 15.
3. $1\frac{5}{7}$ hr.

INEQUALITIES

Modern mathematical thought gives considerable emphasis to the concept of inequality. A meaningful comparison between two quantities can be set up if they are related in some way, even though the relationship may not be one of equality.

The expression "number sentence" is often used to describe a general relationship which may be either an equality or an inequality. If the number sentence states an equality, it is an Equation; if it states an inequality, it is an Inequation.

ORDER PROPERTIES OF REAL NUMBERS

The idea of order, or relative rank according to size, is based upon two intuitive concepts: "greater than" and 'less than."

Mathematicians use the symbol > to represent "greater than" and the symbol < to represent 'less than." For example, the inequation stating that 7 is greater than 5 is written in symbols as follows:

$$7 > 5$$

The inequation stating that x is less than 10 is written as follows:

$$x < 10$$

A "solution" of an inequation involving a variable is any number which may be substi right member.

For example, the inequation $x < 10$ has many solutions. All negative numbers zero, and all positive numbers less than 10, may be substituted for x successfully. These solutions comprise a set of numbers, called the Solution Set.

The Sense of an inequality refers to the direction in which the inequality symbol points. For example, the following two inequalities have opposite sense:

$$7 > 5$$

$$10 < 12$$

PROPERTIES OF INEQUALITIES

Inequations may be manipulated in accordance with specific operational rules, in a manner similar to that used with equations.

Addition

The rule for addition is as follows: If the same quantity is added to both members of an inequation, the result is an inequation having the same sense as the original inequation. The following examples illustrate this:

1. $5 < 8$

   $5 + 2 < 8 + 2$

   $7 < 10$

2. $5 < 8$

$$5 + (-3) < 8 + (-3)$$

$$2 < 50$$

The addition of 2 to both members does not change the sense of the inequation.

The addition of –3 to both members does not change the sense of the inequation. Addition of the same quantity to both members is a useful method for solving inequations.

In the following example, 2 is added to both members in order to isolate the x term on the left:

$$x - 2 > 6$$

$$x - 2 + 2 > 6 + 2$$

$$x > 8$$

Multiplication

The rule for multiplication is as follows: If both members of an inequation are multiplied by the same positive quantity, the sense of the resulting inequation is the same as that of the original inequation. This is illustrated as follows:

1. $-3 < -2$

   $2(-3) < 2(-2)$

   $-6 < -4$

Multiplication of both members by 2 does not change the sense of the inequation.

2. $10 < 12$

   $\frac{1}{2}(10) < \frac{1}{2}(12)$

Multiplication of both members by 1/2 does not change. the sense of the inequation.

Notice that example 2 illustrates division of both members by 2. Since any division can be rewritten as multiplication by a fraction, the multiplication rule is applicable to both multiplication and division. Multiplication is used to simplify the solution of inequations such as the following:

$$\frac{x}{3} > 2$$

Multiply both members by 3:

$$3\left(\frac{x}{3}\right) > 3(2)$$

$$x > 6$$

Sense Reversal

If both sides of an inequation are multiplied or divided by the same negative number, the sense of the resulting inequation is reversed. This is illustrated as follows:

1. $-4 < -2$

   $(-3)(-4) > (-3)(-2)$

   $12 > 6$

2. $7 > 5$

   $(-2)(7) < (-2)(5)$

   $-14 < -10$

Sense reversal is useful in the solution of an inequation in which the variable is preceded by a negative sign, as follows:

$$2 - x < 4$$

Add –2 to both members to isolate the x term:

$$3 = x = 3 < 4 - 2$$

$$-x < 2$$

Multiply both members by – 1:

$$x > -2$$

Practice problems: Solve each of the following inequations:

1. $x + 2 > 3$
2. $\frac{y}{3} - 1 < 2$
3. $3 - x < 6$
4. $4y > 8$

Answers:

1. $x > 1$
2. $y < 9$
3. $x > -3$
4. $y > 2$

GRAPHING INEQUALITIES

An inequation such as $x > 2$ can be graphed on a number line, as shown in figure 11-2. The heavy line in figure contains all values of x which comprise the solution set.

Notice that this line continues indefinitely in the positive direction as indicated by the arrow head. Notice also that the point representing $x = 2$ is designated by a circle. This signifies that the solution set does not contain the number 2.

Figure is a graph of the inequation $x^2 > 4$. Since the square of any number greater than 2 is greater than 4, the solution set contains all values of x greater than 2.

Further less than –2. This is because the square of any negative number smaller than –2 is a positive number greater than 4.

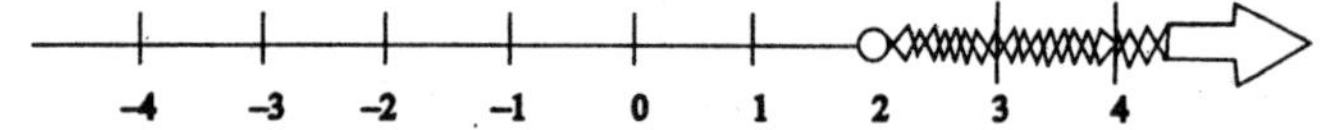

Figure 11.2.- Graph of the in equation $x > 2$.

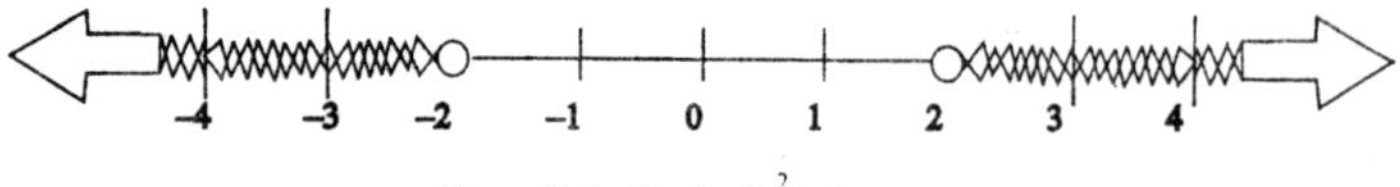

Figure 11.3.- Graph of $x^2 > 4$.

LINEAR EQUATIONS IN TWO VARIABLES

Thus far in this course, discussions of equations have been limited to linear equations in one variable. Linear equations which have two variables are common, and their solution involves extending some of the procedures which have already been introduced.

RECTANGULAR COORDINATES

An outstanding characteristic of equations in two variables is their adaptability to graphical analysis. The rectangular coordinate system, which was introduced in chapter 3 of this course, is used in analyzing equations graphically. This system of vertical and horizontal lines, meeting each other at right angles and thus forming a rectangular grid, is often called the Cartesian coordinate system. It is named after the French philosopher and mathematician, Rene Descartes, who invented it.

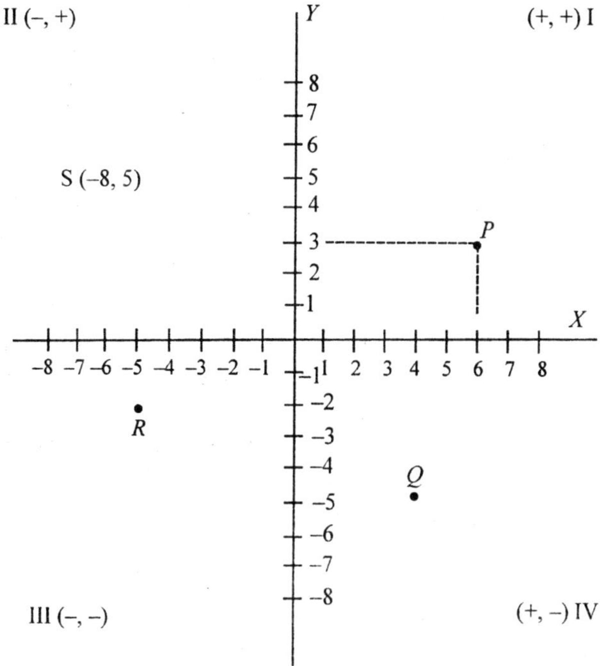

Fig. Rectangular coordinate system. Point P is 6 units to the right of the Y axis and 3 units above the X axis. We call the numbers that indicate the position of a point Coordinates. The number indicating the distance of the point measured horizontally from the origin is the X coordinate (6 in this example), and the number indicating the distance of the point measured vertically from the origin (3 in this example) is the Y coordinate

COORDINATE AXES

The rectangular coordinate system is developed on a framework of reference similar to figure in earlier chapter of this course. On a piece of graph paper, two lines are drawn intersecting each other at right angles.

The vertical line is usually labeled with the capital letter *Y* and called the *Y* axis. The horizontal line is usually labeled with the capital letter *X* and called the *X* axis. The point where the *X* and *Y* axes intersect is called the Origin and is labeled with the letter o. Above the origin, numbers measured along or parallel to the *Y* axis are positive; below the origin they are negative. To the right of the origin, numbers measured along or parallel to the *X* axis are positive; to the left they are negative.

COORDINATES

A point anywhere on the graph may be located by two numbers, one showing the distance of the point from the *Y* axis, and the other showing the distance of the point from the *X* axis.

In describing the location of a point by means of rectangular coordinates, it is customary to place the coordinates within parentheses and separate them with a comma. The *X* coordinate is always written first. The coordinates of point P are written (6, 3). The coordinates for point *Q* are (4, –5); for point *R*, they are (–5, –2); and for point S, they are (–8, 5).

Usually when we indicate a point on a graph, we write a letter and the coordinates of the point. Thus, in figure, for point *S*, we write *S*(–8, 5). The other points would ordinarily be written, *P*(6, 3), *Q*(4, –5), and *R*(–5, –2). The *Y* coordinate of a point is often called its Ordinate and the *X* coordinate is often called its Abscissa.

QUADRANTS

The *X* and *Y* axes divide the graph into four parts called Quadrants. In figure, point *P* is in quadrant I, point *S* is in quadrant II, R is in quadrant III, and Q is in quadrant IV. In the first and fourth quadrants, the X coordinate is positive, because it is to the right of the origin. In the second and third quadrant it is negative, because it is to the left of the origin. Likewise, the Y coordinate is positive in the first and second quadrants, being above the origin; it is negative in the third and fourth quadrants, being below the origin. Thus, we know in advance the signs of the coordinates of a point by knowing the quadrant in which the point appears. The signs of the coordinates in the four quadrants are shown in figure. Locating points with respect to axes is called Plotting. As shown with point P, plotting a point is equivalent to completing a rectangle that has segments of the axes as two of its sides with lines dropped perpendicularly to the axes forming the other two sides. This is the reason for the name "rectangular coordinates."

PLOTTING A LINEAR EQUATION

A linear equation in two variables may have many solutions. For example, in solving the equation $2x - y = 5$, we can find an unlimited number of values of x for which there will be a corresponding value of y. When x is 4, y is 3, since $(2 \times 4) - 3 = 5$. When x is 3, y is 1, and when x is 6, y is 7. When we graph an equation, these pairs of values are considered coordinates of points on the graph. The graph of an equation is nothing more than a line joining the points located by the various pairs of numbers that satisfy the equation.

To picture an equation, we first find several pairs of values that satisfy the equation. For example, for the equation $2x - y = 5$, we assign several values to x and solve for y. A convenient way to find values is to first solve the equation for either variable, as follows:

$$2x - y = 5$$

$$-y = -2x + 5$$

$$y = 2x - 5$$

Once this is accomplished, the value of y is readily apparent when values are substituted for x. The information derived may be recorded in a table such as table 12-1. We then lay off X and Y axes on graph paper, select some convenient unit distance for measurement along the axes, and then plot the pairs of values found for x and y as coordinates of points on the graph. Thus, we locate the pairs of values shown in table on a graph, as shown in figure (A).

Table: Values of x and y in the equation

If x = -------	–2	1	3	5	6	7	8
Then y = —	–9	–3	1	5	7	9	11

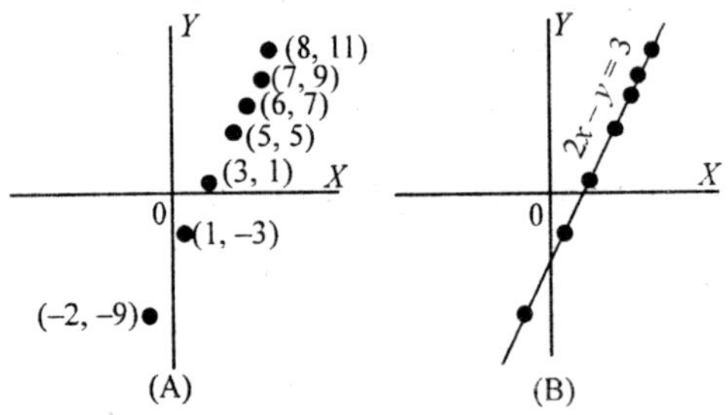

(A) (B)

Finally, we draw a line joining these points, as in figure (B). It is seen that this is a straight line; hence the name "linear equation." Once the graph is drawn, it is customary to write the equation it represents along the line, as shown in figure (B).

It can be shown that the graph of an equation is the geometric representation of all the points whose coo dinates satisfy the conditions of the equation. The line represents an infinite number of pairs of coordinates for this equation. For example, selecting at random the point on the line where x is 2: and y is 0 and substituting these values in the equation, we find that they satisfy it. Thus,

$$2\left(2\frac{1}{2}\right)-0=5$$

If two points that lie on a straight line can be located, the position of the line is known. The mathematical language for this is "Two points Determine a straight line." We now know that the graph of a linear equation in two variables is a straight line.

Since two points are sufficient to determine a straight line, a linear equation can be graphed by plotting two points and drawing a straight line through these points. Very often pairs of whole numbers which satisfy the equation can be found by inspection.

Such points are easily plotted. After the line is drawn through two points, it is well to plot a third point as a check. If this third point whose coordinates satisfy the equation lies on the line the graph is accurately drawn.

X AND *Y* INTERCEPTS

Any straight line which is not parallel to one. of the axes has an *X* intercept and a *Y* intercept. These are the points at which the line crosses the *X* and *Y* axes. At the *X* intercept, the graph line is touching the *X* axis, and thus the *Y* value at that point is 0. At the *Y* intercept, the graph line is touching the *Y* axis; the *X* value at that point is 0.

In order to find the *X* intercept, we simply let $y = 0$ and find the corresponding value of x. The *Y* intercept is found by letting $x = 0$ and finding the corresponding value of y. For example, the line

$$5x + 3y = 15$$

crosses the *Y* axis at (0, 5). This may be verlfied by letting $x = 0$ in the equation. The *X* intercept is (3, 0), since x is 8 when y is 0. Picture shows the Line

$$5x + 3y = 15$$

graphed by means of the *X* and *Y* intercepts.

EQUATIONS IN ONE VARIABLE

An equation containing only one variable its easily graphed, since the line it represents lies parallel to an axis. For example, in

$$2y = 9$$

the value of y is

$$\frac{9}{2}, \text{ or } 4\frac{1}{2}$$

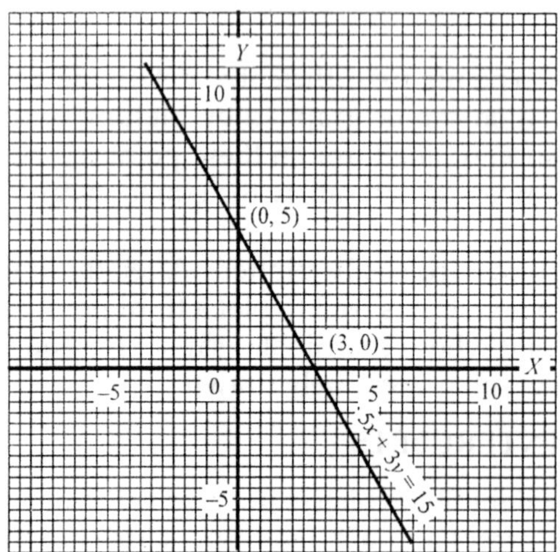

Fig. Graph of $5x + 3y = 15$.

The line $2y = 9$ lies parallel to the X axis at a distance of 4 1/2 units above it. Notice that each small division on the graph paper in figure represents one-half unit.

The line $4x + 15 = 0$ lies parallel to the Y axis. The value of x is – 15/4. Since this value is negative, the line lines to the left of the Y axis at a distance of 3/4 units.

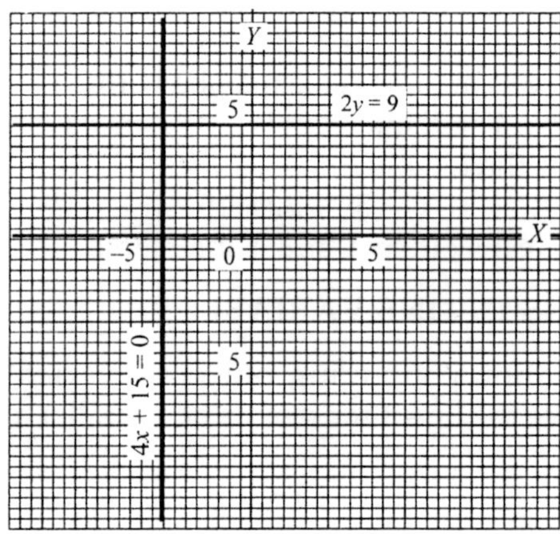

Fig. Graphs of $2y = 9$ and $4x + 15 = 0$.

From the foregoing discussion, we arrive at two important conclusions:

1. A pair of numbers that satisfy an equation are the coordinates of a point on the graph of the equation.
2. The coordinates of any point on the graph of an equation will satisfy that equation.

SOLVING EQUATIONS IN TWO VARIABLES

A solution of a linear equation in two variables consists of a pair of numbers that satisfy the equation. For example, $x = 2$ and $y = 1$ constitute a solution of

$$3x - 5y = 1$$

When 2 is substituted for x and 1 is substituted for y, we have

$$3(2) - 5(1) = 1$$

The numbers $x = -3$ and $y = -2$ also form a solution. This is true because substituting -3 for x and -2 for y reduces the equation to an identity:

$$3(-3) -5(-2) = 1$$

$$-9 + 10 = 1$$

$$1 = 1$$

Each pair of numbers (x, y) such as (2, 1) or (–3, –2) locates a point on the line $3x - 5y = 1$. Many more solutions could be found. Any two numbers that constitute a solution of the equation are the coordinates of a point on the line represented by the equation. Suppose we were asked to solve a problem such as: Find two numbers such that their sum is 35 and their difference is 5. We could indicate the problem algebraically by letting x represent one number and y the other. Thus, the problem may be indicated by the two equations

$$x + y = 33$$

$$x - y = 5$$

Considered separately, each of these equations represents a straight line on a graph. There are many pairs of values for x and y which satisfy the first equation, and many other pairs which satisfy the second equation. Our problem is to find One pair of values that will satisfy Both equations. Such a pair of values is said to satisfy both equations at the same time, or simultaneously. Hence, two equations for which we seek a common solution are called Simultaneous Equations. The two equations, taken together, comprise a System of equations.

Graphical Solution

If there is a pair of numbers that can be substituted for x and y in two different equations, the pair form the coordinates of a point which lies on the graph of each equation. The only way in which a point can lie on two lines simultaneously is for the point to be at the intersection of the lines.

Therefore, the graphical solution of two simultaneous equations involves drawing their graphs and locating the point at which the graph lines intersect. For example, when we graph the equations $x + y = 33$ and $x - y = 5$, as in figure, we see that they intersect in a single point. There is one pair of values comprising coordinates of that point (19, 14), and that pair of values satisfies both equations, as follows:

$$x + y = 33 \qquad\qquad x - y = 5$$

$$19 + 14 = 33 \qquad\qquad 19 - 14 = 5$$

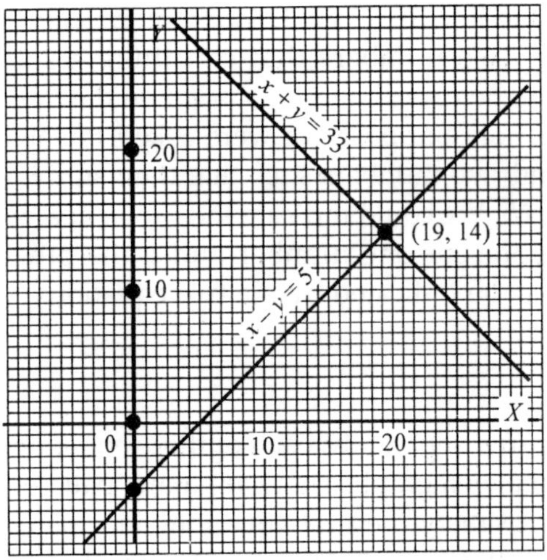

Fig. Graph of $x + y = 33$ and $x - y = 5$.

This pair of numbers satisfies each equation. It is the only pair of numbers that satisfies the two equations simultaneously.

The graphical method is a quick and simple means of finding an approximate solution of two simultaneous equations. Each equation is graphed, and the point of intersection of the two lines is read as accurately as possible.

A high degree of accuracy can be obtained but this, of course, is dependent on the precision with which the lines are graphed and the amount of accuracy possible in reading the graph. Sometimes the graphical method is quite adequate for the purpose of the problem. Figure shows the graphs of $x + y = 11$ and $x - y = -3$. The intersection appears to be the point (4, 7). Substituting $x = 4$ and $y = 7$ into the equations shows that this is the actual point of intersection, since this pair of numbers satisfies both equations.

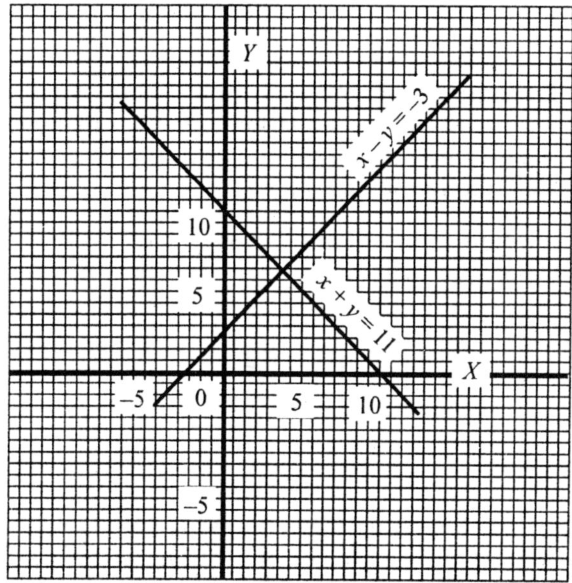

Fig. Graph of $x + y = 11$ and $x - y = -3$

The equations $7x - 8y = 2$ and $4x + 3y = 5$ are graphed in figure. The lines intersect where y is approximately 1/2 and x is approximately 5/6.

Practice problems: Solve the following simultaneous systems graphically:

1. $x + y = 8, x - y = 8$
2. $3x + 2y = 12, 4x + 5y = 2$

Answers:

1. $x = 5, y = 3$
2. $y = 3, y = -6$

Addition Method

The addition method of solving systems of equations is illustrated in the following example:

$$\begin{array}{r} x - y = 2 \\ \underline{x + y = 8} \\ 2x + 0 = 10 \\ x = 5 \end{array}$$

The result in the foregoing example is obtained by adding the left member of the first equation to the left member of the second, and adding the right member of the first equation to the right member of the second.

Having found the value of x, we substitute this value in either of the original equations to find the value of y, as follows:

$$\begin{array}{r} x - y = 2 \\ (5) - y = 2 \\ -y = 2 - 5 \\ -y = -3 \\ y = 3 \end{array}$$

Notice that the primary goal in the addition method is the elimination (temporarily) of,one of the variables. If the coefficient of y is the same in both equations, except for its sign, adding the equations eliminates y as in the foregoing example.

On the other hand, suppose that the coefficient of the variable which we desire to eliminate is exactly the same in both equations. In the following example, the coefficient of x is the same in both equations, including its sign:

$$x + 2y = 4$$
$$x + 3y = -1$$

Adding the equations would not eliminate either x or y. However, if we multiply both members of the second equation by – 1, then addition will eliminate x, as follows:

$$\begin{aligned} x + y &= 4 \\ -x + 3y &= 1 \\ \hline 5y &= 1 \\ y &= 1 \end{aligned}$$

The value of x is found by substituting 1 for y in either of the original equations, as follows:

$$x + 2(1) = 4$$
$$x - 2$$

As a second example of the addition method, find the solution of the simultaneous equations

$$3x + 2y = 12$$
$$4x + 5y = 2$$

Here both x and y have unlike coefficients. The coefficients of one of the variables must be made the same, except for their signs. The coefficients of x will be the same except signs, if both members of the first equation are multiplied by 4 and both members of the second equation by –3. Then addition will eliminate x.

Following this procedure to get the value of y, we multiply the first equation by 4 and the second equation by –3, as follows:

$$\begin{aligned} 12x + 8y &= 48 \\ -12x - 15y &= -6 \\ \hline -7y &= 42 \\ y &= -6 \end{aligned}$$

Substituting for y in the first equation to get the value of x, we have

$$3x + 2(-6) = 12$$
$$\underline{x - 2(-2) = 4}$$
$$x - 4 = 4$$
$$y = 8$$

This solution is checked algebraically by substituting 8 for x and –6 for y in each of the original equations, as follows:

$$3x + 2(-6) = 12$$
$$x + 2(-2) = 4$$
$$x - 4 = 4$$
$$x = 8$$

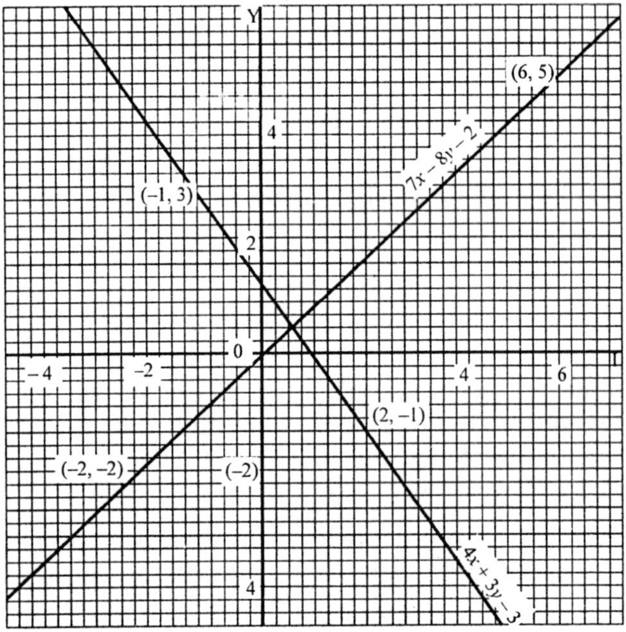

Figure 12-7 Graph of $7x - 8y - 2$ and $4x + 3y - 5$.

1. $3x + 2y = 12$

   $3(8) + 2(-6) = 12$

   $24 - 12 = 12$

2. $4x + 5y = 2$

   $4(8) + 5(-6) = 2$

   $32 - 30 = 2$

Practice problems: Use the addition method to solve the following problems:

1. $x + x = 24, \; x - y = 12$
2. $5t - 2v = 9, \; 3t - 2v = -5$
3. $x - 2y = -1, \; 2x - 3\,y = 12$
4. $2x + 7y + 3, \; 3x - 5y + 51$

Answers:

1. $x = 18, \; y = 6$
2. $t = 1/2, \; v = \frac{13}{4}$
3. $x = 3, \; y = 2$
4. $x = 12, \; y = -3$

Substitution Method

In some cases it is more convenient to use the substitution method of solving problems. In this method we solve one equation for one of the variables and substitute the value obtained into the other equation. This eliminates one of the variables, leaving an equation in one unknown. For example, find the solution of the following system: -.

$$4x + y = 11$$

$$x + 2y = 8$$

It is easy to solve for either y in the first equation or x in the second equation. Let us solve for y in the first equation. The result is

$$y = 11 - 4x$$

Since equals may be substituted for equals, we may substitute this value of y wherever y appears in the second equation. Thus,

$$x + 2(11 - 4x) = 8$$

We now have one equation that is linear in x; that is, the equation contains only the variable x.

Removing the parentheses and solving for x, we find that

$$x + 22 - 22x = 8$$

$$-7x = 8 - 22$$

$$-7x = -14$$

$$x = 2$$

To get the corresponding value of y, we substitute $x = 2$ in $y = 11 - 4x$. The result is

$$y = 11 - 4(2) = 11 - 8 = 3$$

Thus, the solution for the two original equations is $x = 2$ and $y = 9$.

Practice problems: Solve the following systems by the substitution method:

1. $2x - 9x = 1,\ x - 4y = 1$
2. $2x + y = 0,\ 2x - y = 1$
3. $5r - 2s = 29,\ 4r + s = 19$
4. $t - 4v = 1,\ 2t - 9v = 3$

Answers:

1. $x = 5,\ y = 1$
2. $x = 1/4,\ y = -1/2$
3. $r = 3,\ y = 1$
4. $t = -3,\ v = -1$

Literal Coefficients

Simultaneous equations with literal coefficients and literal constants may be solved for the value of the variables just as the other equations discussed in this chapter, with the exception that the solution will contain literal numbers. For example, find the solution of the system:

$$3x + 4y = a$$

$$4x + 3y = b$$

We proceed as with any other simultaneous linear equation. Using the addition method, we may proceed as follows: To eliminate the y term we multiply the first equation by 3 and the second equation by –4. The equations then become

$$\begin{aligned} 9x + 12y &= 3a \\ -16x - 12y &= -4b \\ \hline -7x \qquad &= 3a - 4b \\ x \qquad &= \frac{3a - 4b}{-7} \end{aligned}$$

To eliminate x, we multiply the first equation by 4 and the second equation by –3. The equations then become

We may check in the same manner as that used for other equations, by substituting these values in the original equations.

NTERPRETING EQUATIONS

Recall that the general form for an equation in the first degree in one variable is $ax + b = 0$. The general form for first-degree equations in two variables is

$$ax + by + c = 0$$

It is interesting and often useful to note what happens graphically when equations differ, in certain ways, from the general form. With this information, we know in advance certain facts concerning the equation in question.

LINES PARALLEL TO THE AXES

If in a linear equation the y term is missing, as in

$$2x - 15 = 0$$

the equation represents a line parallel to the Y axis and 7 1/2 units from it. Similarly, an equation such as

$$4y - 9 = 0$$

which has no x term, represents a line parallel to the X axis and 2 1/4 units from it. The fact that one of the two variables does ot appear in an equation means that there are no limitations on the values the missing variable can assume.

When a variable does not appear, it can assume any value from zero to plus or minus infinity. This can happen only if the line represented by the equation lies parallel to the axis of the missing variable.

Lines Passing Through the Origin

A linear equation, such as

$$4x + 3y = 0$$

that has no constant term, represents a line passing through the origin. This fact is obvious since $x = 0$, $y = 0$ satisfies any equation not having a constant term.

Lines Parallel to Each Other

An equation such as

$$3x - 2y = 6$$

has all possible terms present. It represents a line that is not parallel to an axis and does not pass through the origin.

Equations that are exactly alike, except for the constant terms, represent parallel lines. As shown in figure, the lines represented by the equations

$$3x - 2y = -18 \text{ and } 3x - 2y = 6$$

are parallel.

Parallel lines have the same slope. Changing the constant term moves a line away from or toward the origin while its various positions remain parallel to one another. Notice in figure that the line $3x - 2y = 6$ lies closer to the origin than $3x - 2y = -18$. This is revealed at sight for any pair of lines by comparing their constant terms.

That one which has the constant term of greater absolute value will lie farther from the origin. In this case $3x - 2y = -18$ will be farther from the origin since $| -18 | > | 16 |$.

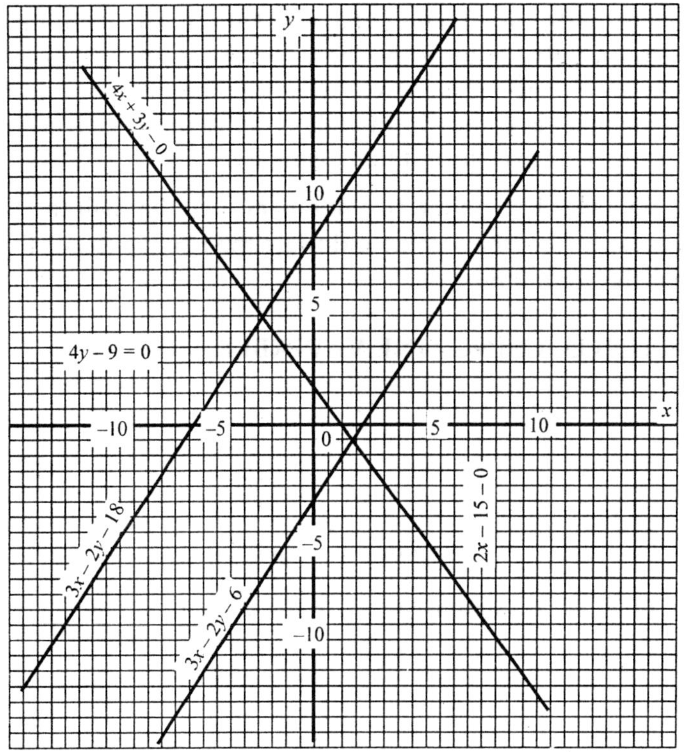

Fig. Interpreting equation

The fact that lines are parallel is indicated by the result when we try to solve two equations such as $3x - 2y = -18$ and $3x - 2y = 6$ simultaneously. Subtraction eliminates both x and y immediately. If both variables disappear, we cannot find values for them such that both equations are satisfied at the same time.

This means that there is no solution. No solution implies that there is no point of intersection for the straight lines represented by the equations. Lines that do not intersect in the finite plane are parallel.

USING TWO VARIABLES IN SOLVING WORD PROBLEMS

Many problems can be solved quickly and easily using one

equation with one variable. Other problems that might be rather difficult to solve in terms of one variable can easily be solved using two equations and two variables. The difference in the two methods is shown in the following example, solved first by using one variable and then using two.

EXAMPLE: Find the two numbers such that half the first equals a third of the second and twice their sum exceeds three times the second by 4.

SOLUTION USING ONE VARIABLE:

1. Let x = the first number
2. Then $\frac{x}{2} = \frac{1}{3}$ of the second number.
3. Thus $\frac{3x}{2}$ = the seond number.

From the statement of the problem, we then have

$$2\left(x + \frac{3x}{2}\right) = 3\left(\frac{3x}{2}\right) + 4$$

$$2x + 3x = \frac{9x}{2} + 4$$

$$10x = 9x + 8$$

$$x = 8 \text{ (first number)}$$

$$\frac{3x}{2} = 12 \text{ (second number)}$$

SOLUTION USING TWO VARIABLES

If we let x and y be the first and second numbers, respectively, we can write two equations almost directly from the statement of the problem. Thus,

1. $\frac{x}{2} = \frac{y}{3}$
2. $2(x + y) = 3y + 4$

Solving for x in the first equation and substituting this value in the second, we have

$$x = \frac{2y}{3},\ 2\left(\frac{2y}{3} + y\right) = 3y + 4$$

$$\frac{4y}{3} + 2y = 3y + 4$$

$$4y + 6y = 9y + 12$$

$$y = 12 \text{ (second number)}$$

$$\frac{x}{2} = \frac{12}{3}$$

$$x = 8 \text{ (first number)}$$

Thus, we see that the solution using two variables is more direct and simple. Often it would require a great deal of skill to manipulate a problem so that it might be solved using one variable; whereas the solution using two variables might be very simple.

The use of two variables, of course, involves the fact that the student must be able to form two equations from the information given in the problem.

Practice problems: Solve the following problems using two variables:

1. A Navy tug averages 12 miles per hour downstream and 9 miles per hour upstream. How fast is the stream flowing?
2. The sum of the ages of two boys is 18. If 4 times the younger boy's age is subtracted from 3 times the older boy's age, the difference is
3. What are the ages of the two boys?

Answers:

1. 1 1/2 mph.
2. 6 years and 12 years.

INEQUALITIES IN TWO VARIABLES

Inequalities in two variables are of the following form:

$$x + y > 2$$

Many solutions of such an inequation are apparent immediately. For example, x could have the value 2 and y could have the value 3, since 2 + 3 is greater than 2.

The existence of a large number of solutions suggests that a graph of the inequation would contain many points. The graph of an inequation in two unknowns is, in fact, an entire area rather than just a line.

PLOTTING ON THE COORDINATE SYSTEM

It would be extremely laborious to plot enough points at random to define an entire area of the coordinate system.

Therefore our method consists of plotting a boundary line and shading the area, on one side of this line, wherein the solution points lie.

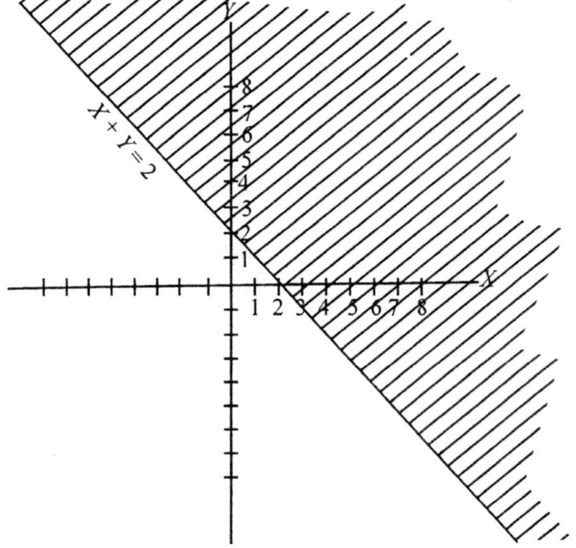

Fig. Graph of $x + y > 2$.

The equation of the boundary line is formed by changing the inequation to an equation. For example, the equation of the boundary line for the graph of

$$x + y > 2$$

is the equation

$$x + y = 2$$

Figure is a graph of $x + y > 2$. Notice that the boundary line $x + y = 2$ is not solid. This is intended to indicate that points on the boundary line are not members of the solution set. Every point lying above and to the right of the boundary line is a member of the solution set. Any solution point may be verified by substituting its X and Y coordinates for x and y in the original inequation.

SIMULTANEOUS INEQUALITIES

The areas representing the solutions of two different inequations may overlap. If such an overlap occurs, the. area of the overlap includes all points whose coordinates satisfy both inequations simultaneously. An example of this is shown in figure, in which the following two inequations are graphed:

$$x + y > 2$$
$$x = y > 2$$

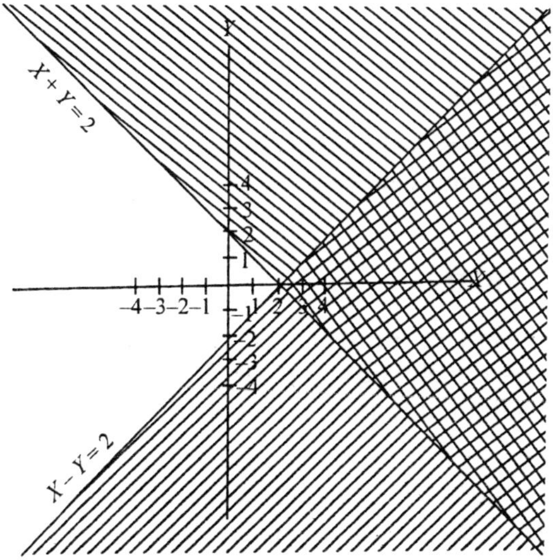

Fig. Graph of $x + y > 2$ and $x - y > 2$.

The double crosshatched area in figure contains all points which comprise the solution set for the system.

Chapter 8

Complex Numbers

In certain calculations in mathematics and related sciences, it is necessary to perform operations with numbers unlike any mentioned thus far in this course. These numbers, unfortunately called "imaginary" numbers by early mathematicians, are quite useful and have a very real meaning in the physical sense.

The number system, which consists of ordinary numbers and imaginary numbers, is called the Complex Number system. Complex numbers are composed of a "real" part and an "imaginary" part.

This chapter is designed to explain imaginary numbers and to show how they can be combined with the numbers we already know.

REAL NUMBERS

The concept of number, as has been noted in previous chapters, has developed gradually. At one time the idea of number was limited to positive whole numbers.

The concept was broadened to include positive fractions; numbers that lie between the whole numbers. At first, fractions included only those numbers which could be expressed with terms that were integers. Since any fraction may be considered as a ratio, this gave rise to the term Rational Number, which is defined as any number which can be expressed as the ratio of two integers. (Remember that any whole number is an integer.)

It soon became apparent that these numbers were not enough to complete the positive number range. The ratio, π, of the circumference of a circle to its diameter, did not fit the concept of number thus far advanced, nor did such numbers as $\sqrt{2}$ and $\sqrt{3}$ Although decimal values are often assigned to these numbers, they are only approximations.

That is, li is not exactly equal to 22/7 or to 3.142. Such numbers are called Irrational to distinguish them from the other numbers of the system. With rational and irrational numbers, the positive number system includes all the numbers from zero to infinity in a positive direction.

Since the number system was not complete with only positive numbers, the system was expanded to include negative numbers.

The idea of negative rational and irrational numbers to minus infinity was an easy extension of the system. Rational and irrational numbers, positive and negative to ± infinity as they have been presented in this course, comprise the Real Number System. The real number system is pictured in figure.

OPERATORS

As shown in a previous chapter, the plus sign in an expression such as 5 + 3 can stand for either of two separate things: It indicates the positive number 3, or it indicates that +3 is to be added to 5; that is, it indicates the op- eration to be performed on +3.

Likewise, in the problem 5 – 3, the minus sign may indicate the negative number –3, in which case the operation would be addition; that is, 5 + (–3). On the other hand, it may indicate the sign of operation, in which case +3 is to be subtracted from 5; that is, 5 – (+3).

Thus, plus and minus signs may indicate positive and negative numbers, or they may indicate operations to be performed.

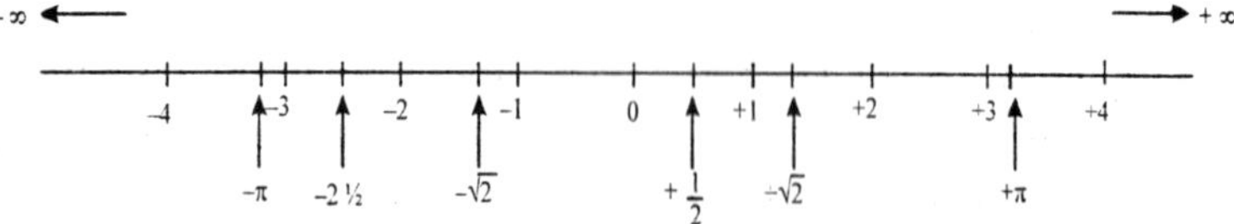

Fig. The Real Number System

IMAGINARY NUMBERS

The number line pictured in figure represents all positive and negative numbers from plus infinity to minus infinity. However, there is a type of number which does not fit into the picture. Such a number occurs when we try to solve the following equation:

$$x^2 + 4 = 0$$

$$x^2 = -4$$

$$x = \pm\sqrt{-4}$$

Notice the distinction between this use of the radical sign and the manner in which it was used in chapter 7. Here, the ± symbol is included with the radical sign to emphasize the fact that two values of x exist. Although both roots exist, only the positive one is usually given. This is in accordance with usual mathematical convention.

The equation raises an interesting question:

$$x = \pm\sqrt{-4}$$

What number multiplied by itself yields –4 ? The square of –2 is +4. Likewise, the square of +2 is +4. There is no number in the system of real numbers that is the square root of a negative number.

The square root of a negative number came to be called an Imaginary Number. When this name was assigned the square roots of negative numbers, it was natural to refer to the other known numbers as the Real Numbers.

IMAGINARY UNIT

To reduce the problem of imaginary numbers to its simplest

terms, we proceed as far as possible using ordinary numbers in the solution. Thus, we may write as a product

$$\sqrt{-1.4} = \sqrt{4}\sqrt{-1} \;= \pm\sqrt{-1}$$

Likewise,

$$3\sqrt{-5} = \sqrt{5}\sqrt{-1}$$

Also,

Thus, the problem of giving meaning to the square root of any negative number reduces to that of finding a meaning for $\sqrt{-1}$

The square root of minus 1 is designated i by mathematicians. When it appears with a coefficient, the symbol i is written last unless the coefficient is in radical form. This convention is illustrated in the following examples:

$$\pm 2\sqrt{-1} = \pm 21$$

$$\sqrt{5}\sqrt{-1} = 1\sqrt{5}$$

$$3\sqrt{7}\sqrt{-1} = 31\sqrt{7}$$

The symbol i stands for the imaginary unit $\sqrt{-1}$. An imaginary number is any real multiple, positive or negative, of i. For example, $-7i$, $+7i$, $i\sqrt{15}$, and bi are all imaginary numbers.

In electrical formulas the letter i denotes current. To avoid confusion, electronic technicians use the letter j to indicate $\sqrt{-1}$, and call it "operator *j*." The name "imaginary" should be thought of as a technical mathematical term of convenience. Such numbers have a very real purpose in the physical sense. Also it can be shown that ordinary mathematical operations such as addition, multiplication, and so forth, may be performed in exactly the same way as for the so-called real numbers.

Practice problems: Express each of the following as some real number times i:

1. $\sqrt{-16}$
2. $2\sqrt{-1}$

3. $\sqrt{-5}$

4. $\frac{d}{f}\sqrt{-f^2}$

5. $\sqrt{-25}$

6. $\sqrt{-\frac{9}{16}}$

Answers:

1. $4i$
2. $2i$
3. $i\sqrt{5}$
4. di
5. $5i$
6. $\frac{3}{4}i$

Powers of the Imaginary Unit

The following examples illustrate the results of raising the imaginary unit to various powers:

$$i^2 = \sqrt{-1}\sqrt{-1}, \text{ or } -1$$
$$i^3 = i^2 i = -1i \text{ or } -i$$
$$i^4 = i^2 i^2 = -1\ .\ -1 = +1$$
$$i^{-1} = \frac{1}{i} = \frac{i}{i^2} = \frac{i}{-1} = -i$$

We see from these examples that an even power of i is a real number equal to +1 or –1. Every odd power of i is imaginary and equal to i or -i. Thus, all powers of i reduce to one of the following four quantities: $\sqrt{-1}, -1, -\sqrt{-1},$ or +1.

GRAPHICAL REPRESENTATION

Figure shows the real numbers represented along a straight

line, the positive numbers extending from zero to the right for an infinite distance, and the negative numbers extending to the left of zero for an infinite distance. Every point on this line corresponds to a real number, and there are no gaps between them. It follows that there is no possibility of representing imaginary numbers on this line.

Earlier, we noted that certain signs could be used as operators. The plus sign could stand for the operation of addition. The minus sign could stand for the operation of subtraction. Likewise, it is easy to explain the imaginary number i graphically as an operator indicating a certain operation is to be performed on the number of which it is the coefficient.

If we graphically represent the length, n, on the number line pictured in figure (A), we start at the point 0 and measure to the right (positive direction) a distance representing n units. If we multiply n by –1, we may represent the result –n by measuring from 0 in a negative direction a distance equal to n units.

Graphically, multiplying a real number by –1 is equivalent to rotating the line that represents the number about the point 0 through 180° so that the new position of n is in the opposite direction and *a* distance n units from 0. In this case we may think of –1 as the operator that rotates n through two right angles to its new position (B)).

As we have shown, $i^2 = -1$. Therefore, we have really multiplied n by i^2, or $i \times i$. In other words, multiplying by –1 is the same as multiplying by i twice in succession. Logically, if we multiplied n by i just once, the line n would be rotated only half as much as before-that is, through only one right angle, or 90°.

The new segment ni would be measured in a direction 90" from the line n. Thus, i is an operator that rotates a number through one right angle.

We have shown previously that a positive number may have two real square roots, one positive and one negative. For example, $\sqrt{9} = \pm 3$.

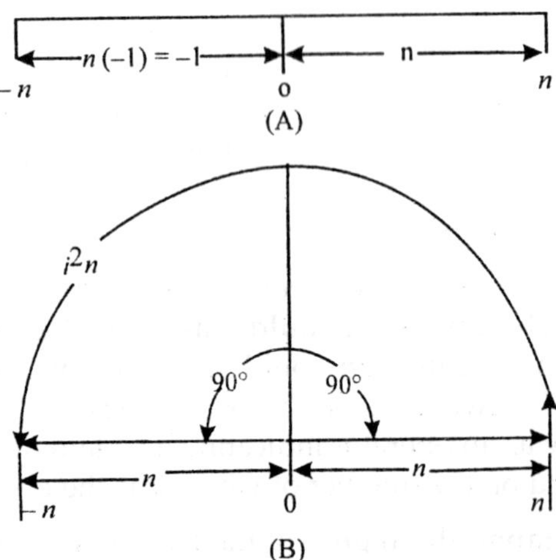

Fig. Graphical Multiplication by −1 and by Operator i^2.

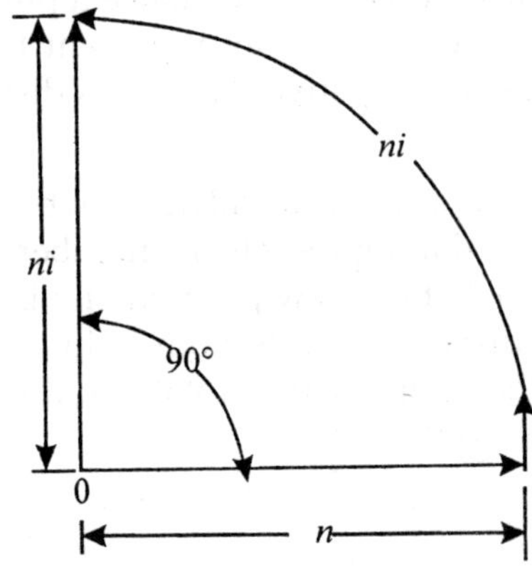

Fig. Graphical Multiplication by Operator *i*.

We also saw that an imaginary number may have two roots. For example, $\sqrt{-4}$ is equal to $\pm 2i$. When the operator −1 graphically rotates a number, it may do so in a counterclockwise or a clockwise direction. Likewise, the operator i may graphically rotate a number in either direction.

This fact gives meaning to numbers such as $\pm 2i$. It has been agreed that a number multiplied by $+i$ is to be rotated 90° in a counterclockwise direction. A number multiplied by $-i$ is to be rotated 90° in a clockwise direction. In figure, $+2i$ is represented by rotating the line that represents the positive real number 2 through 90° in a counterclockwise direction. It follows that $-2i$ is represented by rotating the line that represents the positive real number 2 through 90° in a clockwise direction.

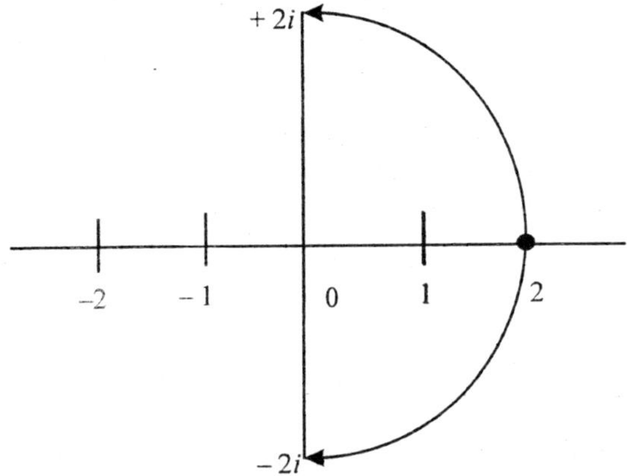

Fig. Graphical Representation of $\pm 2i$.

In figure, notice that the idea of i as an operator agrees with the concept advanced concerning the powers of i. Thus, i rotates a number through 90°; 12 or –1 rotates the number through 180°, and the number is real and negative; i^3 rotates the number through 270°, which has the same effect as $-i$; and i^4 rotates the number through 360°, and the number is once again positive and real.

THE COMPLEX PLANE

All imaginary numbers may be represented graphically along a line extending through zero and perpendicular to the line representing the real numbers. This line may be considered infinite in both the positive and negative directions, and all multiples of i may be represented on it.

This graph is similar to the rectangular coordinate system studied earlier. In this system, the vertical or y axis is called the axis of imaginaries, and the horizontal or x axis is called the axis of reals.

In the rectangular coordinate system, real numbers are laid

off on both the x and y axes and the plane on which the axes lie is called the real plane.

When the y axis is the axis of imaginaries, the plane determined by the x and y axes is called the Complex Plane. In any system of numbers a unit is necessary for counting. Along the real axis, the unit is the number 1.

As shown in figure, along the imaginary axis the unit is i. Numbers that lie along the imaginary axis are called Pure Imaginaries. They will always be some multiple of i, the imaginary unit. The numbers $5i$, $3i$, $\sqrt{2}$ and $\sqrt{-7}$ are examples of pure imaginaries.

NUMBERS IN THE COMPLEX PLANE

All numbers in the complex plane are complex numbers, including reals and pure imaginaries. However, since the reals and imaginaries have the special property of being located on the axes, they are usually identified by their distinguishing names.

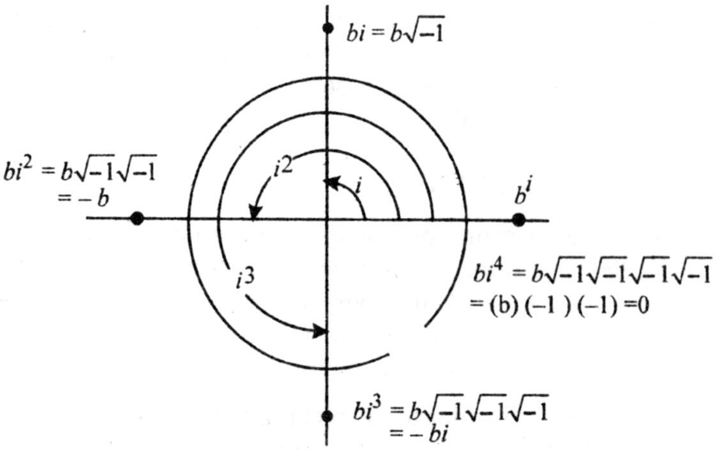

Fig. Operation with Powers of i

The term complex number has been defined as the indicated sum or difference of a real number and an imaginary number.

For example, $3 + 5\sqrt{-1}$ or $3 + 5i$, $2 - 6i$, and $-2 + \sqrt{-5}$ are complex numbers. In the complex number $7 - i$, 7 is the real part and –1 the imaginary part. All complex numbers correspond to the general form $a + bi$, where a and b are real numbers. When a has the value 0, the real term disappears and the complex number becomes a pure imaginary.

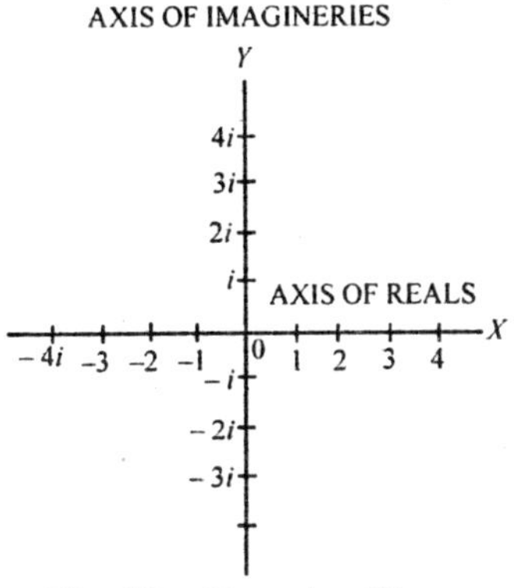

Fig. The Complex Plane.

When b has the value of 0, the imaginary term disappears and the complex number becomes a real number. Thus, 4 may be thought of as $4 + 0i$, and $3i$ may be considered $0 + 3i$. From this we may reason that the real number and the pure imaginary number are special cases of the complex number.

Consequently, the complex number may be thought of as the most general form of a number and can be construed to include all the numbers of algebra as shown in the chart in Figure.

Plotting Complex Numbers

Complex numbers may easily be plotted in the complex plane. Pure imaginaries are plotted along the vertical axis, the axis of imaginaries, and real numbers are plotted along the horizontal axis, the axis of reals. It follows that other points in the complex plane must represent numbers that are part real and part imaginary; in other words, complex numbers. If we wish to plot the point $3 + 2i$, we note that the number is made up of the real number 3 and the imaginary number $2i$.

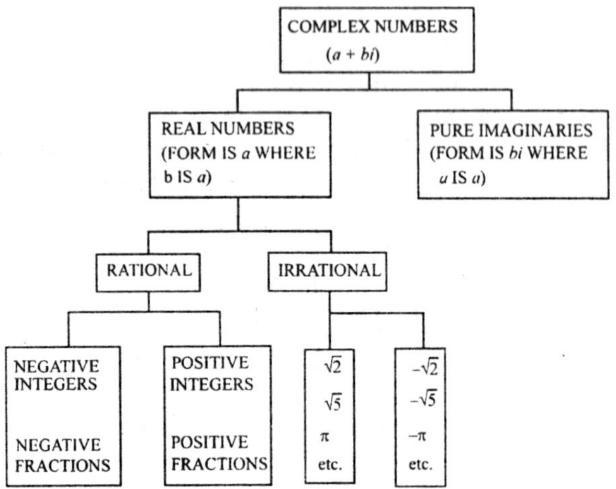

Fig. The Complex Number System.

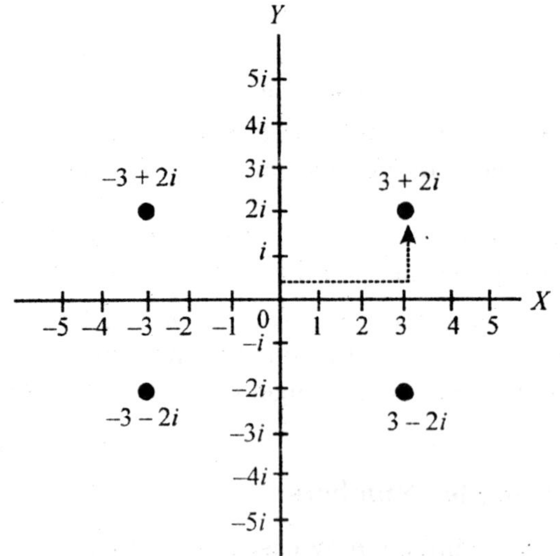

Fig. Plotting Complex Numbers.

Vpositive direction. At point (3, 0) on the real axis we turn through one right angle and measure 2 units up and parallel to the imaginary axis. Likewise, the number $-3 + 2i$ is 3 units to the left and up 2 units; the number $3 - 2i$ is 3 units to the right and down 2 units; and the number $-3 - 2i$ is 3 units to the left and down 2 units.

Complex Numbers as Vectors

A vector is a directed line segment. A complex number represents a vector expressed in the Rectangular Form. For example, the complex number $6 + 8i$ in figure 15 – 9 may be considered as representing either the point P or the line OP. The real parts of the complex number (6 and 8) are the rectangular components of the vector.

The real parts are the legs of the right triangle (sides adjacent to the right angle), and the vector OP is its hypotenuse (side opposite the right angle). If we merely wish to indicate the vector OP, we may do so by writing the complex number that represents it along the segment as in figure. This method not only fixes the position of point P, but also shows what part of the vector is imaginary (PA) and what part is real (OA).

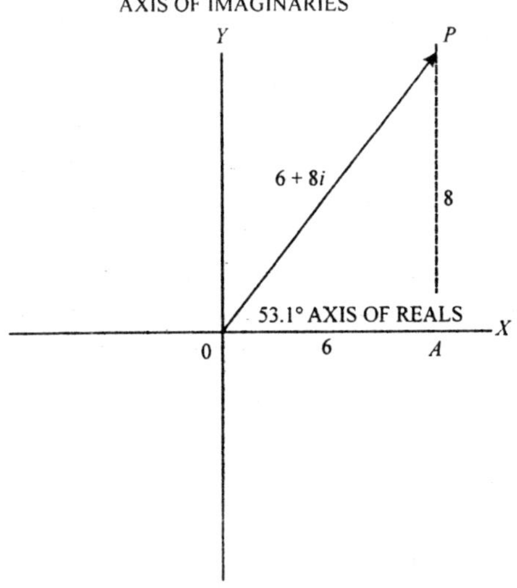

Fig. A Complex number shown as a Vector

If we wish to indicate a number that shows the actual length of the vector OP, it is necessary' to solve the right triangle OAP for its hypotenuse.

This may be accomplished by taking the square root of the sum of the squares of the legs of the triangle, which in this case are the real numbers, 6 and 8. thus,

$$OP = \sqrt{6^1 + 8^2} = \sqrt{100} = 10$$

However, since a vector has direction as well as magnitude, we must also show the direction of the segment; otherwise the segment *OP* could radiate in any direction on the complex plane from point 0.

The expression $10\angle 53.1°$. 10 indicates that the vector *OP* has been rotated counterclockwise from the initial positibn through an angle of 53.1°. (The initial position in a line extending from the origin to the right along *OX*.) This method of expressing the vector quantity is called the Polar Form.

The number represents the magnitude of the quantity, and the angle represents the position of the vector with respect to the horizontal reference, *OX*. Positive angles. represent counterclockwise rotation of the vector, and negative angles represent clockwise rotation. The polar form is generally simpler for multiplication and division, but its use requires a knowledge of trigonometry.

ADDITION AND SUBTRACTION OF COMPLEX NUMBERS

Pure imaginaries are added and subtracted in the same way as any other algebraic quantities. The coefficients of similar terms are added or subtracted algebraically, as follows:

$$4i + 3i = 7i$$

$$4i + 3i = i$$

$$4i + (-3i) = 7i$$

Likewise, complex numbers in the rectangular form are combined like any other algebraic polynomials. Add or subtract the coefficients of similar terms algebraically. If parentheses enclose the numbers, first remove the parentheses. Next, place the real parts together and the imaginary parts together. Collect terms.

As examples, consider the following:

1. $(2 - 3i) + (5 + 4i) = 2 - 3i + 5 + 4i$

$= 2 + 5 - 3i + 4i = 7 + i$

2. $(2 - j3) - (5 + j4) = 2 - j3 - 5 - j4$

$= 2 - 5 - j3 - j4 = -3 - j7$

In example 2, notice that the convention for writing operator j (the electronics form of the imaginary unit) with numerical coefficients is to place j first.

If the complex numbers are placed one under the other, the results of addition and subtraction appear as follows:

Addition	*Subtraction*
$3+4\sqrt{-1}$	$a+jb$
$2-7\sqrt{-1}$	$[\leftarrow c+jd$
$5-3\sqrt{-1}$	$(a-c)+j(b-d)$

Practice problems: Add or subtract as indicated, in the following problems:

1. $(3a + 4i) + (0 - 2i)$
2. $(3 + 2i) + (-3 + 3i)$
3. $(a + bi) + (c + di)$
4. $(1+2\sqrt{-1})+(-2-2\sqrt{-1})$
5. $(-5 + 3i) - (-4 - 2i)$
6. $(a + bi) + (-c + di)$

Answers:

1. $3a + 2i$
2. $5i$
3. $a + c + (b + d)i$
4. -1
5. $-9 + 5i$
6. $a + c + (b - d)i$

MULTIPLICATION OF COMPLEX NUMBERS

Generally, the rules for the multiplication of complex

numbers and pure imaginaries are the same as for other algebraic quantities.

However, there is one exception that should be noted: The rule for multiplying numbers under radical signs does not apply to Two Negative Numbers. When at least one of two radicands is positive, the radicands can be multiplied immediately, as in the following examples:

$$\sqrt{2}\sqrt{3} = \sqrt{6}$$
$$\sqrt{2}\sqrt{-3} = \sqrt{-6}$$

When both radicands are negative, however, as in $\sqrt{-2}\sqrt{-3}$, an inconsistent result is obtained if we multiply both numbers under the radical signs immediately.

To get the correct result, express the imaginary numbers first in terms of i, as follows:

$$\sqrt{-2}\sqrt{-3} = i\sqrt{2}.i\sqrt{3} \; = i^2\sqrt{2}\sqrt{3} \; = i^2\sqrt{6} = (-1)\sqrt{6} = -\sqrt{6}$$

Multiplying complex numbers is equivalent to multiplying binomials in the manner explained previously. After the multiplication is performed, simplify the powers of i as in the following examples:

1. $$\begin{array}{l} 4 - i \\ \underline{3 + i} \\ 12 - 3i \\ \underline{\quad + 4i - i^2} \\ 12 + i - i^2 \; = 12 + i - (-1) \\ \qquad\qquad\quad = 13 + i \end{array}$$

2. $(-6 + 5\sqrt{-7})(8 - 2\sqrt{-7})$

$$= (-6 + 5i\sqrt{7})(8 - 2i\sqrt{7})$$
$$= -48 + 40i\sqrt{7} + 12i\sqrt{7} \; - 10(7)i^2$$

$$= -48 + 52i\sqrt{7} + 70$$

$$= 22 + 52i\sqrt{7}$$

Practice problems: Perform the indicated operations:

1. $\sqrt{-9}\sqrt{-16}$
2. $\sqrt{-2}\sqrt{18}$
3. $\sqrt{-9}\sqrt{-4}$
4. $a\sqrt{ba}.\sqrt{-b}$
5. $(2 + 5i)(3 + 2i)$
6. $(a+\sqrt{-b})(a-\sqrt{-b})$
7. $(-2+\sqrt{-4})(-1+\sqrt{-4})$
8. $(8-\sqrt{-7})(6+\sqrt{-7})$

Answers:

1. -12
2. $6i$
3. -6
4. $-ab\sqrt{a}$
5. $16 + 11i$
6. $a^2 + b$
7. $-2 - 6i$
8. $55 + 21\sqrt{7}$

CONJUGATES AND SPECIAL PRODUCTS

Two complex numbers that are alike except for the sign of their imaginary parts are called Conjugate Complex Numbers. For example, $3 + 5i$ and $3 - 5i$ are conjugates. Either number is the conjugate of the other.

If one complex number is known, the conjugate can be

obtained immediately by changing the sign of the imaginary part. The conjugate of $-8 + \sqrt{-10}$ is $-8\sqrt{-10}$. The conjugate of $-\sqrt{-6}$ is $\sqrt{-6}$.

The sum of two conjugate complex numbers is a real number, as illustrated by the following:

1. $(3 + j5) + (3 - j5) = 2(3) = 6$

2. $\left(-\frac{1}{2}+\frac{\sqrt{-3}}{2}\right)+\left(-\frac{1}{2}-\frac{\sqrt{-3}}{2}\right)$

$$= -\frac{1}{2}+\frac{\sqrt{3}}{2}i-\frac{1}{2}-\frac{\sqrt{3}}{2}i$$

$$= -\frac{1}{2}+\left(-\frac{1}{2}\right) = -1$$

Product of Two Conjugates

The product of two conjugate complex numbers is a real number. Multiplying two conjugates is equivalent to finding the product of the sum and difference of two numbers.

Consider the following examples:

1. $(3 + j5)\,(3 - j5) = 3^2 - (j5)^2 = 9 - 25\,(-1) = 9 + 25 = 34$

2. $\left(-\frac{1}{2}+\frac{\sqrt{3}}{2}i\right)\left(-\frac{1}{2}-\frac{\sqrt{3}}{2}i\right)=\left(-\frac{1}{2}\right)^2-\left(\frac{\sqrt{3}}{2}i\right)^2$

$$= \frac{1}{4}-\left[\frac{3}{4}(-1)\right] = \frac{1}{4}+\frac{3}{4} = 1$$

Squaring a Complex Number

Squaring a complex number is equivalent to raising a binomial to the second power. For example:

$$(-6-\sqrt{-25})^2 = (-6 - j5)^2$$

$$= [(-1) \times (6 + j5)]^2$$

$$= (-1)^2 \times (6^2 + j60 + j^2 25)$$

$$= 36 + j60 - 25 = 11 + j60$$

DIVISION OF COMPLEX NUMBER

When dividing by a pure imaginary, the denomfnator may be rationalized and the problem thus simplified by multiplying both numerator and denominator by the denominator. Thus,

$$\frac{12}{\sqrt{-2}} = \frac{12}{i\sqrt{2}} \cdot \frac{i\sqrt{2}}{i\sqrt{2}} = \frac{12i\sqrt{2}}{2i^2} = \frac{6i\sqrt{2}}{-1} = -6i\sqrt{2}$$

Division of complex numbers can be accomplished by multiplying the numerator and denominator by the number that is the conjugate of the denominator. This process is similar to the process of rationalizing a denominator in the case of real numbers that are irrational.

As an example, consider

$$\frac{5-2i}{3+i}$$

The denominator is $3 + i$. Its conjugate is $3 - i$. Multiplying numerator and denominator by $3 - i$ gives

$$\frac{5-2i}{3+i} \cdot \frac{3-i}{3-i} = \frac{15-11i+2i^2}{9-i^2} = \frac{15-11i-2}{9+i}$$

$$= \frac{13-11i}{10} = \frac{13}{10} - \frac{11}{10}i$$

Practice problems: Rationalize the denominators and simplify:

1. $\dfrac{2\sqrt{-1}}{4+2\sqrt{-1}}$

2. $\dfrac{-2+4i}{-1+4i}$

3. $\dfrac{3+\sqrt{-2}}{3-\sqrt{-2}}$

4. $\dfrac{3}{1-i\sqrt{3}}$

5. $\dfrac{1-i}{2-i}$

6. $\dfrac{8}{2+\sqrt{-2}}$

Answers:

1. $\dfrac{2i+1}{5}$

2. $\dfrac{18+4i}{17}$

3. $\dfrac{7+6i\sqrt{2}}{11}$

4. $\dfrac{3}{4}+\dfrac{3}{4}i\sqrt{3}$

5. $\dfrac{3-i}{5}$

6. $\dfrac{8-4i\sqrt{2}}{2}$

Chapter 9

Quadratic Equations

The degree of an equation in one variable is the exponent of the highest power to which the variable is raised in that equation. A second degree equation in one variable is one in which the variable is raised to the second power.

A second-degree equation is often called a Quadratic Equation. The word quadratic is derived from the Latin word quadratus, which means "squared." In a quadratic equation the term of highest degree is the squared term. For example, the following are quadratic equations:

$$x^2 + 3x + 4 = 0$$

$$3m + 4m^2 = 6$$

The terms of degree lower than the second may or may not be present. The possible terms of lower degree than the squared term in a quadratic equation are the first-degree term and the constant term. In the equation

$$3x^2 - 8x - 5 = 0$$

-5 is the coefficient of x^0. If we wished to emphasize the powers of x in this equation, we could write the equation in the form

$$3x^2 - 8x^1 - 5x^0 = 0$$

Examples of quadratic equations in which either the first-degree term or the constant term is missing are:

1. $4x^2 = 16$

2. $y^2 + 16y = 0$
3. $c^2 + 12 = 0$

GENERAL FORM OF A QUADRATIC EQUATION

Any quadratic equation can be arranged in the general form:

$$ax^2 + bx + c = 0$$

If it has more than three terms, some of them will be alike and can be combined, after which the final form will have at most three terms. For example,

$$2x^2 + 3 + 5x - 1 + x^2 = 4 - x^2 - 2x - 3$$

reduces to the simpler form

$$4x^2 + 7x + 1 = 0$$

In this form, it is easy to see that a, the coefficient of x^2, is 4; b, the coefficient of x, is 7; and c, the constant term, is 1.

Sometimes the coefficients of the terms of a quadratic appear as negative numbers, as follows:

$$2x^2 - 3x - 5 = 0$$

This equation can be rewritten in such a way that the connecting signs are all positive, as in the general form. This is illustrated as follows:

$$2x^2 + (-3)x + (-5) = 0$$

In this form, the value of a is seen to be 2, b is –3, and c is –5.

An equation of the form

$$x^2 + 2 = 0$$

has no x term. This can be considered as a case in which a is 1 (coefficient of x^2 understood to be l), b is 0, and c is 2. For the purpose of emphasizing the values of a, b, and c with reference to the general form, this equation can be written

$$x^2 + 0x + 2 = 0$$

The coefficient of x^2 can never be 0; if it were 0, the equation would not be a quadratic. If the coefficients of x and x^0 are 0, then those terms do not normally appear. To say that the coefficient of x^0 is 0 is the same as saying that the constant term is 0 or is missing.

A Root of an equation in one variable is a value of the variable that satisfied the equation. Every equation in one variable, with constants as coefficients and positive integers as exponents, has as many roots as the exponent of the highest power. In other words, the number of roots is the same as the degree of the equation. A fourth-degree equation has four roots, a cubic (third-degree) equation has three roots, a quadratic equation has two roots, and a linear equation has one root.

As an example, 6 and – 1 are roots of the quadratic equation

$$x^2 - 5x - 6 = 0$$

This can be verified by substituting these values into the equation and noting that an identity results. in each case.

Substituting $x = 6$ gives

$$6^3 - 5(6) - 6 = 0$$

$$36 - 36 = 0$$

$$0 = 0$$

Substituting $x = -1$ gives

$$(-1)^2 - 5(-1) - 6 = 0$$

$$1 + 5 - 6 = 0$$

$$6 - 6 = 0$$

$$0 = 0$$

Several methods of finding the roots of quadratic equations (Solving) are possible. The most common methods are solution by Factoring and solution by the Quadratic Formula. Less commonly used methods of solution are accomplished by completing the square and by graphing.

SOLUTION BY FACTORING

The equation $x^2 - 36 = 0$ is a pure quadratic equation, There are two numbers which, when substituted for x, will satisfy the equation as follows:

$$(+6)^2 - 36 = 0$$
$$36 - 36 = 0$$

also

$$(-6)^2 - 36 = 0$$
$$36 - 36 = 0$$

Thus, +6 and – 6 are roots of the equation

$$x^2 - 36 = 0$$

The most direct way to solve a pure quadratic (one in which no x term appears and the constant term is a perfect square) involves rewriting with the constant term in the right member, as follows:

$$X^2 = 36$$

Taking square roots on both sides, we have

$$x = \pm 6$$

The reason for expressing the solution as both plus and minus 6 is found in the fact that both +6 and –6, when squared, produce 36.

The equation

$$x^2 - 36 = 0$$

can also be solved by factoring, as follows:

$$x^2 - 36 = 0$$
$$(x + 6)(x - 6) = 0$$

We now have the product of two factors equal to zero. According to the zero factor law, if a product is zero, then one or more of its factors is zero. Therefore, at least one of the factors must be zero, and it makes no difference which one. We are free to set first one factor and then the other factor equal to zero. In so doing we derive two solutions or roots of the equation. If $x + 6$ is the factor whose value is 0, then we have

$$x + 6 = 0$$

$$x = -6$$

If $x - 6$ is the zero factor, we have

$$x - 6 = 0$$

$$x = 6$$

When a three-term quadratic is put into simplest form, it is customary to place all the terms on the left side of the equality sign with the squared term first, the-first-degree term next, and the constant term last, as in

$$9x^2 - 2x + 7 = 0$$

If the trinomial in the left member is readily factorable, the equation can be solved quickly by separating the trinomial into factors. Consider the equation

$$3x^2 - x - 2 = 0$$

By factoring the trinomial, the equation becomes

$$(3x + 2)(x - 1) = 0$$

Once again we have two factors, the product of which is 0. This means that one or the other of them (or both) must have the value 0. If the zero factor is $3x + 2$, we have

$$3x + 2 = 0$$

$$3x = -2$$

$$x = -\frac{2}{3}$$

If the zero factor is $x - 1$, we have

$$x - 1 = 0$$

$$x = 1$$

Substituting first $x = 1$ and then $x = -2/3$ in the original equation, we see that both roots satisfy it. Thus,

$$3(1)^2 - (1) - 2 = 0$$

$$3 - 1 - 2 = 0$$

$$0 = 0$$

$$3\left[-\frac{2}{3}\right]^2 - \left[-\frac{2}{3}\right] - 2 = 0$$

$$\frac{4}{3}+\frac{2}{3}-2=0$$

$$0=0$$

In summation, when a quadratic may be readily factored, the process for finding its roots is as follows:

1. Arrange the equation in the order of the descending powers of the variable so that all the terms appear in the left member and zero appears in the right.
2. Factor the left member of the equation.
3. Set each factor containing the variable equal to zero and solve the resulting equations.
4. Check by substituting each of the derived roots in the original equation.

Example: Solve the equation $x^2 - 4x = 12$ for x.

1. $x^2 - 4x - 12 = 0$
2. $(x - 6)(x + 6) = 0$
3. $x - 6 = 0 \qquad x + 2 = 0$

   $x = 6 \qquad x = -2$
4. $(6)^2 -4(8) = 12 \ (x = 6)$

   $36 - 24 = 12$

   $12 = 12$

   $(-2)^2 -4(-2) = 12 \ (x = -2)$

   $4 + 8 = 12$

   $12 = 12$

Practice problems: Solve the following equations by factoring:

1. $x^2 + 10x - 24 = 0$
2. $a^2 - a - 56 = 0$
3. $y^2 - 2y - 63$
4. $y^2 - 19y - 6 = 0$
5. $m^2 - 4m = 96$

Answers:

1. $x = -12, x = 2$
2. $a = 8, a = -7$
3. $y = -7, y = 9$
4. $y = 3, y = -\frac{2}{7}$
5. $m = -8, m = 12$

SOLUTION BY COMPLETING THE SQUARE

When a quadratic cannot be solved by factoring, or the factors are not readily seen, another method of finding the roots is needed. A method that may always be used for quadratics in one variable involves perfect square trinomials. These, we recall, are trinomials whose factors are identical. For example,

$$x^2 - 10x + 25 = (x - 5)(x - 5) = (x - 5)^2$$

Recall that in squaring a binomial, the third term of the resulting perfect square trinomial is always the square of the second term of the binomial. The coefficient of the middle term of the trinomial is always twice the second term of the binomial. For example, when $(x + 4)$ is squared, we have

$$\begin{array}{l} x+4 \\ \underline{x+4} \\ x^2+4x \\ \underline{\quad +4x+16} \\ x^2+8x+16 \end{array}$$

Hence if both the second- and first-degree terms of a perfect square trinomial are known, the third may be written by squaring one-half the coefficient of the first-degree term. Essentially, in completing the square, certain quantities are added to one member and subtracted from the other, and the equation is so arranged that the left member is a perfect square trinomial. The square roots of both members may then be taken, and the subsequent equalities may be solved for the variable.

For example,

$$x^2 + 5x - \frac{11}{4} = 0,$$

Cannot be readily factored. To solve for x by completing the square, we proceed as follows:

1. Leave only the second- and first-degree terms in the left member.

   $$x^2 + 5x = \frac{11}{4}$$

   (If the coefficient of x^2 is not 1, divide through by the coefficient of x^2.)

2. Complete the square by adding to both members the square of half the coefficient of the x term. In this example, one-half of the coefficient of the x term is 5/2, and the square of 5/2 is 25/4. Thus,

   $$x^2 + 5x + \frac{25}{4} = \frac{11}{4} + \frac{25}{4}$$

3. Factor the left member and simplify the right member.

   $$\left(x + \frac{5}{2}\right)^2 = 9$$

4. Take the square root of both members.

   $$\sqrt{\left(x + \frac{5}{2}\right)^2} = \pm\sqrt{9}$$

   Remember that, in taking square roots on both sides of an equation, we must allow for the fact that two roots exist in every second-degree equation. Thus we designate both the plus and the minus root of 9 in this example.

5. Solve the resulting equations.

   $$x + \frac{5}{2} = 3 \qquad x + \frac{5}{2} = -3$$

$$x = \frac{6}{2} - \frac{5}{2} \qquad x = -\frac{6}{2} - \frac{5}{2}$$

$$x = \frac{1}{2} \qquad x = -\frac{11}{2}$$

6. Check the results.

$$\left(\frac{1}{2}\right)^2 + \frac{5}{2} - \frac{11}{4} = 0$$

$$\frac{5}{2} - \frac{10}{4} = 0$$

$$0 = 0$$

$$\left(-\frac{11}{2}\right)^2 + (5)\left(-\frac{11}{2}\right) - \frac{11}{4} = 0$$

$$\frac{121}{4} - \frac{55}{2} - \frac{11}{4} = 0$$

$$0 = 0$$

The process of completing the square may always be used to solve a quadratic equation. However, since this process may become complicated in more complex equations, a formula based on completing the square has been developed in which known quantities may be substituted in order to derive the roots of the quadratic equation. This formula is explained in the following paragraphs.

SOLUTION BY THE QUADRATIC FORMULA

The quadratic formula is derived by applying the process of completing the square to solve for x in the general form of the quadratic equation, $ax^2 + bx + c = 0$. Remember that the general form represents every possible quadratic equation. Thus, if we can solve this equation for x, the solution will be in terms of a, b, and c. To solve this equation for x by completing the square, we proceed as follows:

1. Subtract the constant term, c, from both members.

$$ax^2 + bx + c = 0$$

$$ax^2 + bx = -c$$

2. Divide all terms by a so that the coefficient of the x^2 term becomes unity.

$$x^2 + \frac{b}{a}x = -\frac{c}{a}$$

3. Add the square of one-half the coefficient of the x term b/a, to both members.

$$\text{Square}\frac{b}{2a}:\left(\frac{b}{2a}\right)^2 = \frac{b^2}{4a^2}$$

$$\text{Add: } x^2 + \frac{b}{a}x + \frac{b^2}{4a^2} = \frac{b^2}{4a^2} - \frac{c}{a}$$

4. Factor the left member and simplify the right member.

$$\left(x + \frac{b}{2a}\right)^2 = \frac{b^2 - 4ac}{4a^2}$$

5. Take the square root of both members.

$$x + \frac{b}{2a} = \pm\frac{\sqrt{b^2 - 4ac}}{4a}$$

6. Solve for x.

$$x = -\frac{b}{2a} \pm \frac{\sqrt{b^2 - 4ac}}{2a} = \frac{-b \pm \sqrt{b^2 - 4ac}}{2a}$$

Thus, we have solved the equation representing every quadratic for its unknown in terms of its constants a, b, and c. Hence, in a given quadratic we need only substitute in the expression

$$\frac{-b \pm \sqrt{b^2 - 4ac}}{2a}$$

the values of a, b, and c, as they appear in the particular equation, to derive the roots of that equation. This expression is called the Quadratic Formula.

The general quadratic equation, $ax^2 + bx + c = 0$, and the quadratic formula should be memorized. Then, when a quadratic cannot be solved quickly by factoring, it may be solved at once by the formula.

Example: Use the quadratic formula to solve the equation

$$x^2 + 30 - 11x = 0.$$

Solution:

1. Set up the equation in standard form.

   $x^2 - 11x + 30 = 0$

   Then a (coefficient of x^2) = 1

   b (coefficient of x) = –11

   c (the constant term) = 30

2. Substituting,

$$x = \frac{-b \pm \sqrt{b^2 - 4ac}}{2a} = \frac{-(-11) \pm \sqrt{(-11)^2 - 4(1)(30)}}{2(1)}$$

$$= \frac{11 \pm \sqrt{121 - 120}}{2} = \frac{11 \pm 1}{2} = 6 \text{ or } 5$$

3. Checking:

When	When
$x = 6,$	$x = 5,$
$(6)^2 - 11(6) + 30 = 0$	$(5)^2 - 11(6) + 30 = 0$
$36 - 66 + 30 = 0$	$25 - 55 + 30 = 0$
$0 = 0$	$0 = 0$

Example: Find the roots of

$$2x^2 - 3x - 1 = 0$$

Here, $a = 2$, $b = -3$, and $c = -1$.

Substituting into the quadratic formula gives

$$x = \frac{-(-3) \pm \sqrt{(-3)^2 - (2)(-1)}}{2(2)} = \frac{3 \pm \sqrt{9 + 8}}{4}$$

$$= \frac{3 \pm \sqrt{17}}{4}$$

The two roots are

$$x = \frac{3}{4} + \frac{1}{4}\sqrt{17} \text{ and } x = \frac{3}{4} - \frac{1}{4}\sqrt{17}$$

These roots are irrational numbers, since the radicals cannot be removed. If the decimal values of the roots are desired, the value of the square root of 17 can be taken from appendix I of this course. Substituting $\sqrt{17} = 4.1231$ and simplifying gives

$$x_1 = \frac{3 + 4.1231}{4} \text{ and } x_2 = \frac{3 - 4.1231}{4}$$

$$x_1 = \frac{7.1231}{4},\ x_2 = \frac{-1.1231}{4}$$

$x_1 = 1.781$, $x_2 = -0.281$

In decimal form, the roots of $2x^2 - 3x - 1 = 0$ to the nearest tenth are 1.8 and –0.3.

Notice that the subscripts, 1 and 2, are used to distinguish between the two roots of the equation. The three roots of a cubic equation in x might be designated x_1, x_2, and x_1. Sometimes the letter r is used for root. Using r, the roots of a cubic equation could be labeled r_1, r_2, and r_3.

Checking:

When $\quad x_1 = \dfrac{3 + \sqrt{17}}{4}$

$$2x^2 - 3x - 1 = 0$$

then

$$2\left(\frac{3 + \sqrt{17}}{4}\right)^2 - 3\left(\frac{3 + \sqrt{17}}{4}\right) - 1 = 0$$

$$\left(\frac{3 + \sqrt{17}}{8}\right)^2 - \frac{9 + 3 + \sqrt{17}}{4} - 1 = 0$$

$$\frac{9+6\sqrt{17}+17-18-6\sqrt{17}-8}{8}=0$$

$$0 = 0$$

When

$$x_2=\frac{3-\sqrt{17}}{4}$$

then

$$2\left(\frac{3-\sqrt{17}}{4}\right)^2-3\left(\frac{3-\sqrt{17}}{4}\right)-1=0$$

$$\frac{9-6\sqrt{17}+17}{8}-\frac{9-3\sqrt{17}}{4}-1=0$$

Multiplying both members of the equation by 8, the LCD, we have

$$8\left(\frac{9-6\sqrt{17}+17}{8}\right)-8\left(\frac{9-3\sqrt{17}}{4}\right)-8(1)=0$$

$$9-6\sqrt{17}+17-2(9-3\sqrt{17})-8=0$$

$$9-6\sqrt{17}+17-18+6\sqrt{17})-8=0$$

$$0 = 0$$

Practice problems: Use the quadratic formula to find the roots of the following equations:

1. $3x^2 -20 - 7x = 0$
2. $4x^2 - 3x - 5 = 0$
3. $15x^2 - 22x -5 = 0$
4. $x^2 + 7x = 8$

Answers:

1. $x_1 = 4,\ x_2 = -\frac{5}{3}$
2. $x_1=\frac{3+\sqrt{89}}{8}$, $x_2=\frac{3-\sqrt{89}}{8}$

3. $x_1 = \frac{5}{3}, \; x_2 = -\frac{1}{5}$

4. $x_1 = 1, x_2 = -8$

GRAPHICAL SOLUTION

A fourth method of solving a quadratic equation is by means of graphing. In graphing linear equations using both axes as reference, we recall that an independent variable, x, and a dependent variable, y, were needed. The coordinates of points on the graph of the equation were designated (x, y).

Since the quadratics we are considering contain only one variable, as in the equation

$$x^2 - 8x + 12 = 0$$

We cannot plot values for the equations in the present form using both x and y axes. A dependent variable, y, is necessary. If we think of the expression

$$x^2 - 8x + 12$$

as a function, then this function can be considered to have many possible numerical values, depending on what value we asign to x. The particular value or values of x which cause the value of the function to be 0 are solutions for the equation

$$x^2 - 8x + 12 = 0$$

For convenience, we may choose to let y represent the function

$$x^2 - 8x + 12$$

If numerical values are now assigned to x, the corresponding values of y may be calculated. When these pairs of corresponding values of x and y are tabulated, the resulting table provides the information necessary for plotting a graph of the function.

Example: Graph the equation

$$x^2 + 2x - 8 = 0$$

and from the graph write the roots of the equation.

Solution:

1. Let $y = x^2 + 2x - 8$.
2. Make a table of the y values corresponding to the value assigned x, as shown in table .

Table: Tabulation of x and y values for the function $y = x^2 + 2x - 8$

if $x = ...$	-5	-4	-3	-2	-1	0	1	2	3
then $y...$	7	0	-5	-8	-9	-8	-5	0	7

3. Plot the pairs of x and y values that appear in the table as coordinates of points on a rectangular coordinate system as in figure (A).
4. Draw a smooth curve through these points, as shown in figure (B).

Notice that this curve crosses the X axis in two places. We also recall that, for any point on the X axis, the y coordinate is zero. Thus, in the figure we see that when y is zero, x is -4 or $+2$. When y is zero, furthermore, we have the original equation,

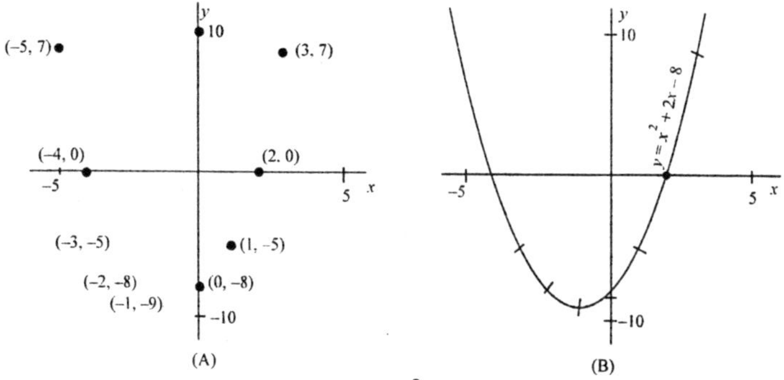

Fig. Graph of the Equation $y = x^2 + 2x - 8$. (A) Points Plotted; (B) Curve drawn through Plotted points

$$x^2 + 2x - 8 = 0$$

Thus, the values of x at these points where the graph of the equation crosses the X axis ($x = -4$ or $+2$) are solutions to the original equation. We may check these results by solving the equation algebraically. Thus,

$$x^2 + 2x - 8 = 0$$

$$(x + 4)(x - 2) = 0$$

$$x_1 + 4 = 0 \qquad x_2 - 2 = 0$$

$$x_1 = -4 \qquad x_2 = 2$$

Check:

$$(-4)^2 + 2(-4) - 8 = 0 \qquad (2)^2 + 2(2) - 8 = 0$$

$$16 - 8 - 8 = 0 \qquad 4 + 4 - 8 = 0$$

$$0 = 0 \qquad 0 = 0$$

The curve in figure (B) is called a Parabola. Every quadratic of the form $ax^2 + bx + c = y$ will have a graph of this general shape. The curve will open downward if a is negative, and upward if a is positive.

Graphing provides a fourth method of finding the roots of a quadratic in one variable. When the equation is graphed, the roots will be the X intercepts (those values of x where the curve crosses the X axis). The X intercepts are the points at which y is 0.

MAXIMUM AND MINIMUM POINTS

It will be seen from the graphs of quadratics in one variable that a parabola has a maximum or minimum value, depending on whether the curve opens upward or downward. Thus, when a is negative the curve passes through a maximum value; and when a is positive, the curve passes through a minimum value. Often these maximum or minimum values comprise the only information needed for a particular problem.

In higher mathematics it can be shown that the X coordinate, or abscissa, of the maximum or minimum value is

$$x = \frac{-b}{2a}$$

In other words, if we divide minus the coefficient of the x term by twice the coefficient of the x^2 term, we have the X coordinate of the maximum or minimum point. If we substitute this value for x in the original equation, the result is the Y value or ordinate, which corresponds to the X value.

For example, we know that the graph of the equation

$$x^2 + 2x - 8 = y$$

passes through a minimum value because a is positive. To find the coordinates of the point where the parabola has its minimum value, we note that $a = 1$, $b = 2$, $c = -8$. From the rule given above, the X value of the minimum point is

$$x = \frac{-b}{2a}$$

$$x = \frac{-(2)}{2(1)}$$

$$x = -1$$

Substituting this value for x in the original equation, we have the value of the Y coordinate of the minimum point. Thus,

$$(-1)^2 + 2(-1) - 8 = y$$

$$-9 = y$$

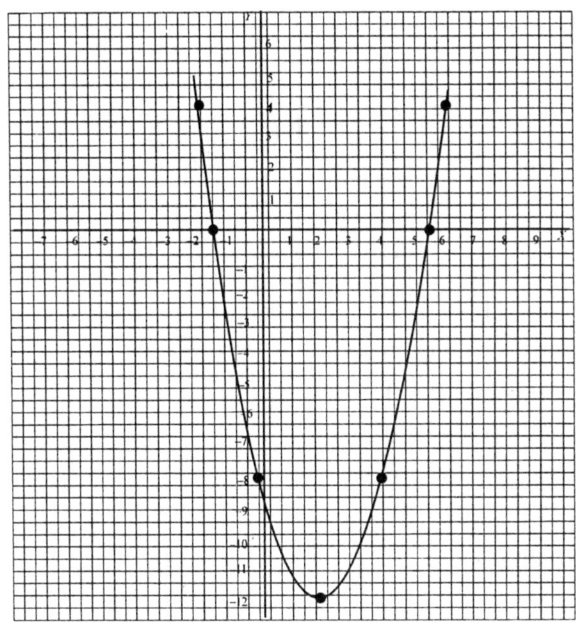

Fig. Graph of $x^2 - 4x - 8 = 0$

The minimum point is (–1, –9). From the graph in figure (A), we see that these coordinates are correct. Thus, we can quickly and easily find the coordinates of the minimum or maximum point for any quadratic of the form $ax^2 + bx + c = 0$.

Practice problems: Without graphing, find the coordinates of the maximum or minimum points for the following equations and state whether they are maximum or minimum.

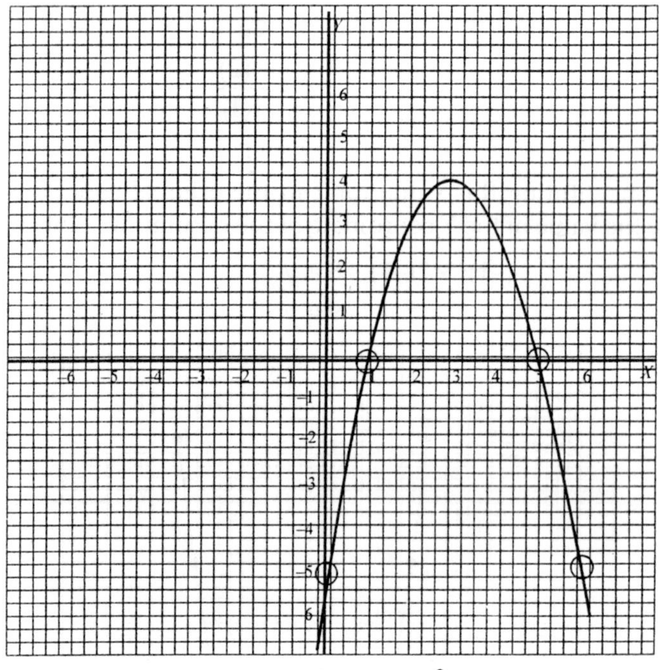

Fig. Graph of $6x - 5 - x^2 = 0$

1. $2x^2 - 2x + 2 = 0$
2. $68 - 3x - x^2 = 0$
3. $3 + 7x - 6x^2 = 0$
4. $24x^2 - 14x = 3$

Answers:

1. $x = \frac{5}{4}, \ y = -\frac{9}{8}$ Minimum
2. $x = -\frac{3}{2}, \ y = \frac{281}{4}$ Maximum

3. $x = \frac{7}{12}, \; y = \frac{121}{24}$ Maximum

4. $x = \frac{7}{24}, \; y = -\frac{121}{24}$ Minimum

THE DISCRIMINANT

The roots of a quadratic equation may be classified in accordance with the following criteria:

1. Real or imaginary.
2. Rational or irrational.
3. EquaI or unequal.

The task of discriminating among these possible characteristics to find the nature of the roots is best accomplished with the aid of the quadratic formula. The part of the quadratic formula which is used is called the Discriminant. If the roots of a quadratic are denoted by the symbols r_1 and r_2, then the following relations may be stated:

$$r_1 = \frac{-b + \sqrt{b^2 - 4ac}}{2a}$$

$$r_2 = \frac{-b - \sqrt{b^2 - 4ac}}{2a}$$

We can show that the character of the roots is dependent upon the form taken by the expression

$$b^2 - 4ac$$

which is the quantity under the radical in the formula. This expression is the Disciuminant of a quadratic equation.

IMAGINARY ROOTS

Since there is a radical in each root, there is a possibility that the roots could be imaginary. They are imaginary when the number under the radical in the quadratic formula is negative (less than 0). In other words, when the value of the discriminant Is less than 0, the roots are imaginary.

Example:

$$x^2 + x + 1 = 0$$

$$a = 1, b = 1, c = 1$$

$$b^2 - 4ac = (1)^2 - 4(1)(1) = 1 - 4 = -3$$

Thus, without further work, we know that the roots are imaginary.

Check: The roots are

$$r_1 = \frac{-1+\sqrt{-3}}{2}, \; r_2 = \frac{-1-\sqrt{-3}}{2}$$

$$r_1 = -\frac{1}{2}+\frac{i\sqrt{3}}{2}, \; r_2 = -\frac{1}{2}-\frac{i\sqrt{3}}{2}$$

We recognize both of these numbers as being imaginary.

We may also conclude that when one root is imaginary the other will also be imaginary. This is because the pairs of imaginary roots are always conjugate complex numbers. If one root is of the form a + ib, then a - ib is also a root. Knowing that imaginary roots always occur in pairs, we can conclude that a quadratic equation always has either two imaginary roots or two real roots.

Practice problems. Using the discriminant, state whether the roots of the following equations are real or imaginary:

1. $x^2 - 6x - 16 = 0$
2. $x^2 - 6x = -12$
3. $3x^2 - 10x + 50 = 0$
4. $6x^2 + x = 1$

Answers:

1. Real
2. Imaginary
3. Imaginary
4. Real

EQUAL OR DOUBLE ROOTS

If the discriminant $b^2 - 4ac$ equals zero, the radical in the quadratic formula becomes zero. In this case the roots are equal; such roots are sometimes called double roots.

Consider the equation

$$9x^2 + 12x + 4 = 0$$

Comparing with the general quadratic, we notice that

$$a = 9,\ b = 12,\ \text{and}\ c = 4$$

The discriminant is

$$b^2 - 4ac = 12^2 - 4(9)(4) = 144 - 144 = 0$$

Therefore, the roots are equal.

Check: From the formula

$$r_1 = -\frac{-12+0}{2(9)},\ r_2 = -\frac{-12-0}{2(9)}$$

$$r_1 = -\frac{2}{3},\ r_2 = -\frac{2}{3}$$

The equality of the roots is thus verified.

The roots can be equal only if the trinomial is a. perfect square. Its factors are equal. Factoring the trinomial in

$$9x^2 + 12x + 4 = 0$$

we see that

$$(3x + 2)^2 = 0$$

Since the factor $3x + 2$ is squared, we actually have

$$3x + 2 = 0$$

twice, and we have

$$x = -\frac{2}{3}$$

twice.

The fact that the same root must be counted twice explains the use of the term "double root." A double root of a quadratic equation is always rational because a double root can occur only when the radical vanishes.

REAL AND UNEQUAL ROOTS

When the discriminant is positive, the roots must be real. Also they must be unequal since equal roots occur only when the discriminant is zero.

Rational Roots

If the discriminant is a perfect square, the roots are rational. For example, consider the equation

$$3x^2 - x - 2 = 0$$

in which $a = 3, b = -1, \text{ and } c = -2$

The discriminant is

$$b^2 - 4ac = (-1)^2 - 4(3)(-2) = 1 + 24 = 25$$

We see that the discriminant, 25, is a perfect square. The perfect square indicates that the radical in the quadratic formula can be removed, that the roots of the equation are rational, and that the trinomial can be factored. In other words, when we evaluate the discriminant and find it to be a perfect square, we know that the trinomial can be factored.

Thus,

$$3x^2 - x - 2 = 0$$

$$(3x + 2)(x - 1) = 0$$

from which

$$3x + 2 = 0, x = -\frac{2}{3}$$

$$x - 1 = 0, x = 1$$

We see that the information derived from the discriminant is correct. The roots are real, unequal, and rational.

Irrational Roots

If the discriminant is not a perfect square, the radical cannot be removed and the roots are irrational.

Consider the equation

$$2x^2 - 4x + 1 = 0$$

in which

$$a = 2, b = -4, \text{ and } c = 1.$$

The discriminant is

$$b^2 - 4ac = (-4)^2 - 4(2)(1) = 16 - 8 = 8$$

This discriminant is positive and not a perfect square. Thus the roots are real, unequal, and irrational.

To check the correctness of this information, we derive the' roots by means of the formula. Thus,

$$x = \frac{-b \pm \sqrt{b^2 - 4ac}}{2a} = \frac{4 \pm \sqrt{8}}{4} = \frac{2 \pm \sqrt{2}}{2}$$

$$x = 1 + \frac{\sqrt{2}}{2} \text{ or } x = 1 - \frac{\sqrt{2}}{2}$$

This verifies the conclusions reached in evaluating the discriminant. When the discriminant is a positive number, not a perfect square, it is useless to attempt to factor the trinomial. The formula is needed to find the roots. They will be real, unequal, and irrational.

SUMMARY

The foregoing information concerning the discriminant may be summed up in the following four rules:

1. If $b^2 - 4ac$ is a perfect square or zero, the roots are rational; otherwise they are irrational.
2. If $b^2 - 4ac$ is negative (less than zero), the roots are imaginary.
3. If $b^2 - 4ac$ is zero, the roots are real, equal, and rational.
4. If $b^2 - 4ac$ is greater than zero, the roots are real and unequal.

Practice problems: Determine the character of the roots of each of the following equations:

1. $x^2 - 7x + 12 = 0$

2. $9x^2 - 6x + 1 = 0$
3. $2x^2 - x + 1 = 0$
4. $9x - 2x^2 + 6 = 0$

Answers:

1. Real, unequal, rational
2. Real, equal, rational
3. Imaginary
4. Real, unequal, irrational

GRAPHICAL INTERPRETATION OF ROOTS

When a quadratic is set equal to y and the resulting equation is graphed, the graph will reveal the character of the roots, but it may not reveal whether the roots are rational or irrational.

Consider the following equations:

1. $x^2 + 6x - 3 = y$
2. $x^2 + 6x + 9 = y$
3. $x^2 + 6x + 13 = y$

The graphs representing these equations are shown in figure 16-4.

We recall that the roots of the equation are the values of x at those points where y is zero. Y is zero on the graph anywhere along the X axis.

Thus, the roots of the equation are the positions where the graph crosses the X axis. In parabola No. 1 we see immediately that there are two roots to the equation and that they are unequal. These roots appear to be –6.5 and 0.5. Algebraically, we find them to be the irrational numbers

$$-3 + 2\sqrt{3} \text{ and } -3 - 2\sqrt{3}.$$

For equation No. 2, the parabola just touches the X axis at $x = -3$. This means that both roots of the equation are the same that is, the root is a double root. At the point where the parabola touches the X axis, the two roots of the quadratic equation have moved together and the two points of intersection of the parabola and the X axis are coincident.

The quantity –3 as a double root agrees with the algebraic solution. When the equation Number is solved algebraically, we see that the roots are $-3 + 2i$ and $-3 - 2i$.

Thus they are imaginary. Parabola No. 3 does not cross the X axis. When this situation occurs, imaginary roots are implied.

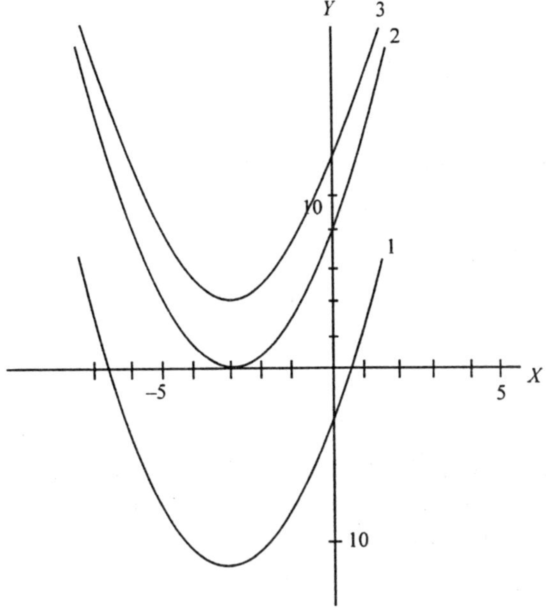

Fig. Graphical Interpretation of Roots.

Only equations having real roots will have graphs that cross or touch the X axis. Thus we may determine from the graph of an equation whether the roots are real or imaginary.

VERBAL PROBLEMS

INVOLVING QUADRATIC EQUATIONS

Many practical problems give rise to quadratic equations. In such problems it often happens that one of the roots will have no meaning. We must select the root that satisfies the conditions of the problem.

Consider the following example: The length of a plot of ground exceeds its width by 7 ft and the area of the plot is 120 sq ft. What are the dimensions?

Solution: Let x = length y = width

then

$$x - y = 7$$

and

$$xy = 120$$

Solving for y, $y = x - 7$

Substituting $(x - 7)$ for y

$$x(x - 7) = 120$$

Therefore

$$x^2 - 7x - 120 = 0$$

$$(x - 15)(x + 8) = 0$$

$$x = 15, x = -8$$

Thus, length = +15 or –8.

But the length obviously cannot be a negative value. Therefore, we reject –8 as a value, for x and use only the positive value, +15. Then from equation

$$15 - y = 7$$

$$y = 8$$

Length = 15, width = 8

Practice problems: Solve the following problems by forming quadratic equations:

1. A rectangular plot is 8 yd by 24 yd. If the length and width are increased by the same amount, the area is increased by 144 sq yd. How much is each dimension increased?
2. Two cars travel at uniform rates of speed over the

same route a distance of 180 ml. One goes 5 mph slower than the other and takes 1/2 hr longer to make the run. How fast does each car travel?

Answers:

1. Length and width are each increased by 4yd.
2. Faster car: 45 mph.

   Slower car: 40 mph.

Chapter 10

Numerical Trigonometry

The word "trigonometry" means "measurement by triangles." As it is presented in many textbooks, trigonometry includes topics other than triangles and measurement. However, this chapter is intended only as an introduction to the numerical aspects of trigonometry as they relate to measurement of lengths and angles.

SPECIAL PROPERTIES OF RIGHT TRIANGLES

A Right Triangle has been defined as any triangle containing a right angle. The side opposite the right angle in a right triangle is a Hypotenuse. In figure, side *AC* is the hypotenuse.

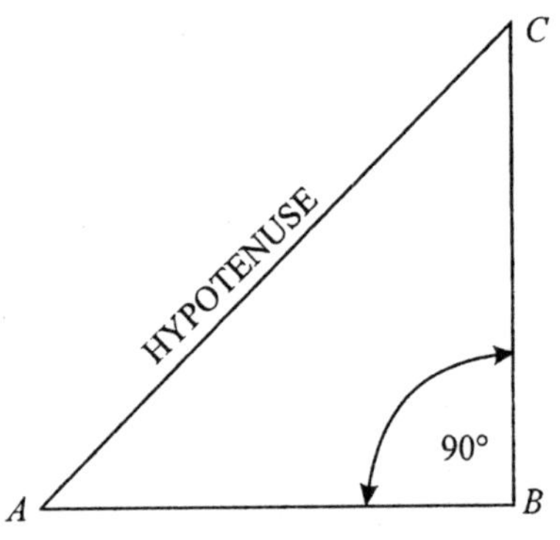

Fig. A Right Triangle.

An important property of all right triangles, which relates

the lengths of the three sides, was discovered by the Greek philosopher Pythagoras.

PYTHAGOREAN THEOREM

The rule of Pythagoras, or Pythagoras Theorem, states that the square of the length of the hypotenuse (in any right triangle) is equal to the sum of the squares of the lengths of the other two sides. For example, if the sides are

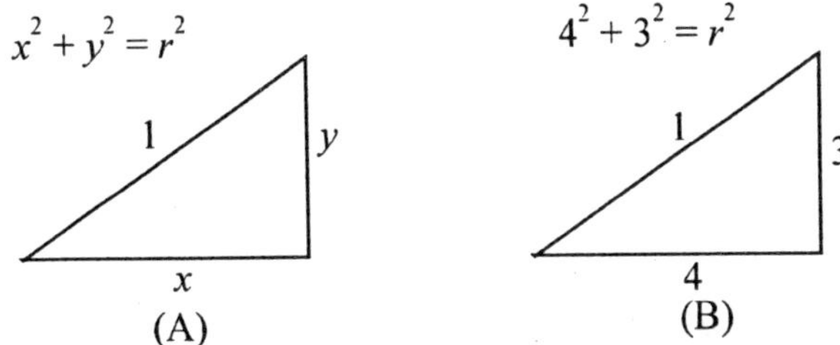

Fig. The Pythagorean Theorem. (A) General Triangle; (B) Triangle with sides of Specific Lengths. Labeled as in figure (A), the Pythagorean

Theorem is stated in symbols as follows:

$$x^2 + y^2 = r^2$$

An example of the. use of the Pythagorean Theorem in a problem follows:

Example: Find the length of the hypotenuse in the triangle shown in figure (B).

Solution:

$$r^2 = 3^2 + 4^2$$

$$r = \sqrt{9+16} = \sqrt{25} = 5$$

Example: An observer on a ship at point *A*, figure, knows that his distance from point *C* is 1,200 yards and that the length of *BC* is 1,300 yards. He measures angle *A* and finds that it is 90°. Calculate the distance from *A* to *B*.

Solution: By the rule of Pythagoras,

$$(BC)^2 = (AB)^2 + (AC)^2$$

$$(BC)^2 = (AB)^2 + (1{,}200)^2$$

$$(1{,}300)^2 - (1{,}200)^2 = (AB)^2$$

$$(13 \times 10^2)^2 - (12 \times 10^2)^2 = (AB)^2$$
$$(169 \times 10^4) - (144 \times 10^4) = (AB)^2$$
$$25 \times 10^4 = (AB)^2$$
$$500 \text{ yd} = AB$$

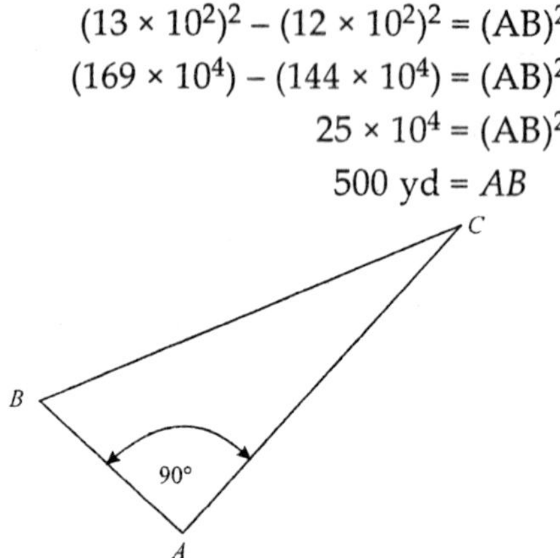

Fig. Using the Pythagorean Theorem

SIMILAR RIGHT TRIANGLES

Two right triangles are SIMILAR if one of the acute angles of the first is equal to one of the acute angles of the second. This conclusion is supported by the following reasons:

1. The right angle in the first triangle is equal to the right angle in the second, since all right angles are equal.
2. The sum of the angles of any triangle is 180°. Therefore, the sum of the two acute angles in a right triangle is 90°.
3. Let the equal acute angles in the two triangles be represented by A and A′ respectively. Then the other acute angles, B and B′, are as follows:

$$B = 90° - A$$
$$B' = 90° - A'$$

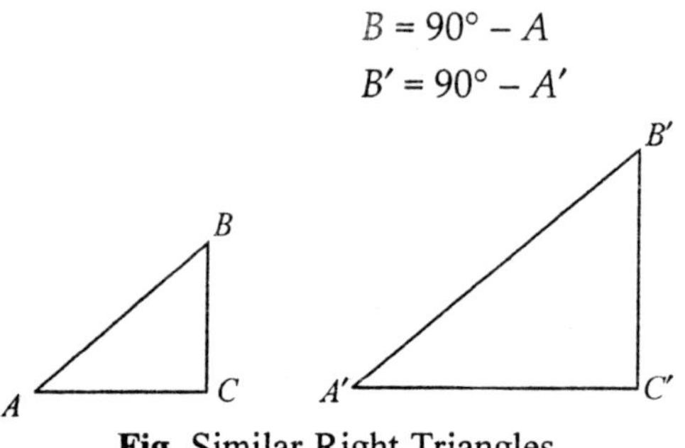

Fig. Similar Right Triangles

4. Since angles A and A' are equal, angles B and B' are also equal.
5. We conclude that two right triangles with one acute angle of the first equal to one acute angle of the second have all of their corresponding angles equal. Thus the two triangles are similar.

Practical situations frequently occur in which similar right triangles are used to solve problems. For example, the height of a tree can be determined by comparing the length of its shadow with that of a nearby flagpole, as shown in figure.

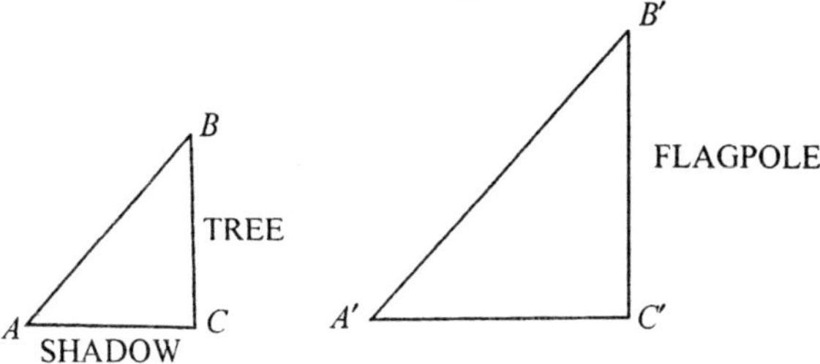

Fig. Calculation of Height by Comparison of Shadows

Assume that the rays of the sun are parallel and that the tree and flagpole both form 90° angles with the ground. Then triangles ABC and $A'B'C'$ are right triangles and angle B is equal to angle B'.

Therefore, the triangles are similar and their corresponding sides are proportional, with the following result:

$$\frac{BC}{AC} = \frac{B'C'}{A'C}$$

$$BC = \frac{(AC)\times(B'C')}{A'C'}$$

Suppose that the flagpole is known to be 30 feet high, the shadow of the tree is 12 feet long, and the shadow of the flagpole is 24 feet long. Then

$$BC = \frac{12\times 30}{24} = \text{feet}$$

Practice Problems:

1. A mast at the top of a building casts a shadow whose tip is 48 feet from the base of the building. If the building is 12 feet high and its shadow is 32 feet long, what is the length of the mast? (NOTE: If the length of the mast is x, then the height of the mast above the ground is $x + 12$.)
2. Figure 19-6 represents an L-shaped building with dimensions as shown. On the line of sight from A to D, a stake is driven at C, a point 8 feet from the building and 10 feet from A. If ABC is a right angle, find the length of AB and the length of AD. Notice that AE is 18 feet and ED is 24 feet.

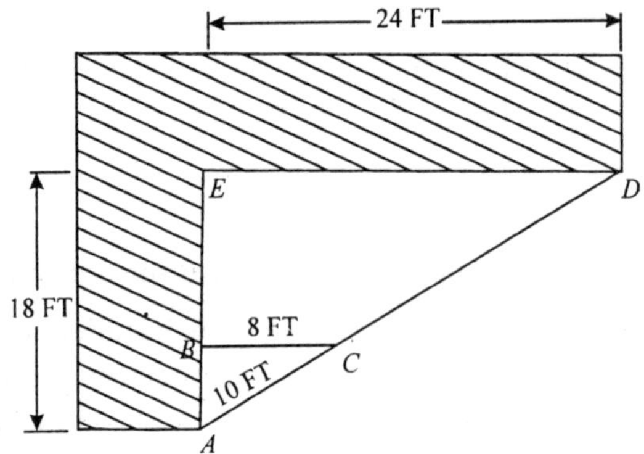

Fig. Using Similar Triangles

Answers:

1. 6 feet
2. $AB = 6$ feet

   $AD = 30$ feet

TRIGONOMETRIC RATIOS

The relationships between the angles and the sides of a right triangle are expressed in terms of Trigonometric Ratios. For example, in figure, the sides of the triangle are named in accordance with their relationship to angle q. In trigonometry, angles are usually named by means of Greek letters. The Greek name of the symbol q is theta.

The six trigonometric ratios for the angle q are listed in table.

The-ratios are defined as follows:

1. $\cos\,\theta = \dfrac{\text{side opposite}\,\theta}{\text{hypotenuse}} = \dfrac{y}{r}$

2. $\cos\,\theta = \dfrac{\text{side adjacnet to}\,\theta}{\text{hypotenuse}} = \dfrac{x}{r}$

3. $\tan\,\theta = \dfrac{\text{side opposite}\,\theta}{\text{side adjacent to}\,\theta} = \dfrac{y}{x}$

4. $\cot\,\theta = \dfrac{\text{side adjacent to}\,\theta}{\text{side opposite to}\,\theta} = \dfrac{x}{y}$

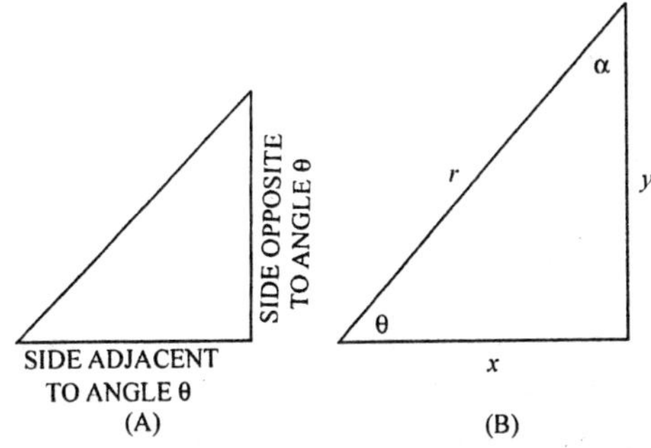

Fig. Relationship of Sides and Angles in a right Triangle. (A) Names of the Sides;
(B) Symbols used to Designate the Sides

Table: Trigonometric ratios

Name of ratio	Abbreviation
sine of θ	sin θ
cosine of θ	cos θ
tangent of θ	tan θ
cotangent of θ	cot θ
secant of θ	sec θ
cosecant of θ	csc θ

5. $\csc\theta = \dfrac{\text{hypotenuse}}{\text{side opposite to }\theta} = \dfrac{r}{x}$

6. $\csc\theta = \dfrac{\text{hypotenuse}}{\text{side opposite to }\theta} = \dfrac{r}{y}$

The other acute angle in figure (B) is labeled a (Greek alpha). The side opposite a is x and the side adjacent to a is y. Therefore the six ratios for a are as follows:

1. $\sin\alpha = \dfrac{x}{r}$
2. $\cos\alpha = \dfrac{y}{r}$
3. $\tan\alpha = \dfrac{x}{y}$
4. $\cot\alpha = \dfrac{y}{x}$
5. $\sec\alpha = \dfrac{r}{y}$
6. $\csc\alpha = \dfrac{r}{x}$

Suppose that the sides of triangle (B) in figure are as follows: $x = 3$, $y = 4$, $r = 5$. Then each of the ratios for angles q and a may be expressed as a common fraction or as a decimal. For example,

$$\sin\theta = \frac{4}{5} = 0.800$$

$$\sin\alpha = \frac{3}{5} = 0.600$$

Decimal values have been computed for ratios of angles between 0° and 90°, and values for angles above 90" can be expressed in terms of these same values by means of conversion formulas. Appendix II of this training course gives the sine, cosine, and tangent of angles from 0° to 90°. The secant, cosecant, and cotangent are calculated,' when needed, by using their relationships to the three principal ratios. These relationships are as follows:

$$\text{secant } \theta = \frac{1}{\text{cosine } \theta}$$

$$\text{cosecant } \theta = \frac{1}{\sin\theta}$$

$$\text{cosecant } \theta = \frac{1}{y \text{ tangent } \theta}$$

TABLES

Tables of decimal values for the trigonometric ratios may be constructed in a variety of ways. Some give the angles in degrees, minutes, and seconds; others in degrees and tenths of a degree.

The latter method is more compact and is the method used for appendix II. The "headings" at the bottom of each page in appendix II provide a convenient reference showing the minute equivalents for the decimal fractions of a degree. For example, 12′ (12 minutes) is the equivalent of 0.2°.

Finding the Function Value

The trigonometric ratios are sometimes called Functions, because the value of the ratio depends upon (is a function of) the angle size. Finding the function value in appendix II is easily accomplished. For example, the sine 35° is found by looking in the "sin" row opposite the large number 35, which is located in the extreme left-hand column.

Since our angle in this example is exactly 35°, we look for the decimal value of -the sine in the column with the 0.0° heading. This column contains decimal values for functions of the angle plus 0.0°; in our example, 35° plus 0.0°, or simply 35.0°. Thus we find that the sine of 35.0° is 0.5736. By the same reasoning, the sine of 42.7° is 0.6782, and the tangent of 32.3° is 0.6322.

A typical problem in trigonometry is to find the value of an unknown side in a right triangle when only one side and one acute angle are known. Example. In triangle *ABC*, find the length of *AC* if *AB* is 13 units long and angle *CAB* is 34.7°.

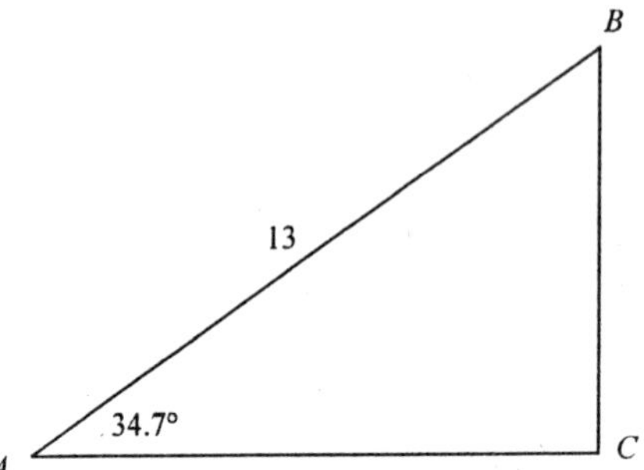

Fig. Using the Trigonometric Ratios to Evaluate the Sides

Solution:

$$\frac{AC}{13} = \cos 34.7°$$

$$AC = 13 \cos 34.7°$$

$$= 13 \times 0.8221$$

$$= 10.69 \text{ (approx.)}$$

The angles of a triangle are frequently stated in degrees and minutes, rather than degrees and tenths. For example, in the foregoing problem, the angle might have been stated as 34° 42′.

When the stated number of minutes is an exact multiple of 6 minutes, the minute entries at the bottom of each page in appendix II may be used.

Finding the Angle

Problems are frequently encountered in which two sides are known, in a right triangle, but neither of the acute angles is known. For example, by applying the Pythagorean Theorem we can verify that the triangle in figure 19-9 is a right triangle, The only information given, concerning angle q, is the ratio of sides in the triangle. The size of q is calculated as follows:

$$\tan \theta = \frac{5}{12} = 0.4167$$

θ = the angle whose tnagent is 0.4167

Assuming that the sides and angles in figure 19-9 are in

approximately the correct proportions, we estimate that angle *q* is about 20°.

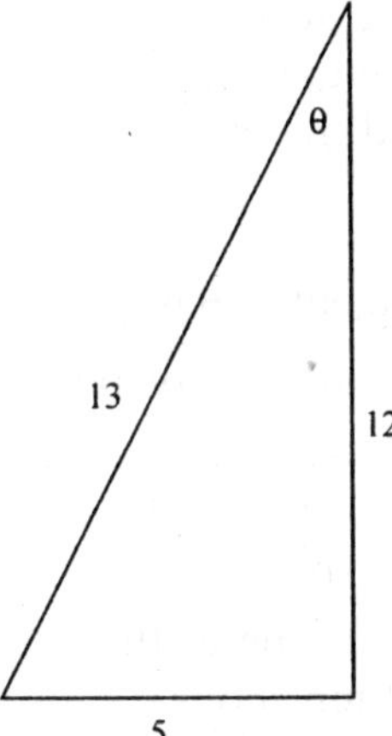

Fig. Using Trigonometric Ratios to Evaluate Angles

The table entries for the tangent in the vicinity of 20 are slightly too small, since we need a number near 0.4167. However, the tangent of 22° 36′ is 0.4163 and the tangent of 22° 42′ is 0.4183. Therefore, q is. between 22° 36′ and 22° 42′.

Interpolation

It is frequently necessary to estimate the value of an angle to a closer approximation than is available in the table. This is equivalent to estimating between table entries, and the process is called Interpolation. For example in the foregoing problem it was determined that the angle value was between 22° 36′ and 22° 42′. The following paragraphs describe the procedure for interpolating to find a closer approximation to the value of the angle.

The following arrangement of numbers is recommended for interpolation:

$$\begin{array}{ccc} & \textit{Angle} & \textit{Tangent} \\ 6' & \left[\begin{array}{l} 22°36' \\ \theta \\ 22°42' \end{array}\right. & \left.\left.\begin{array}{r} 0.4163 \\ 0.4167 \end{array}\right\} .0004 \atop 0.4183 \right] .0020 \end{array}$$

The spread between 22° 36′ and 22° 42′ is 6′, and we use the comparison of the tangent values to determine how much of this 6′ spread is included in q, the angle whose value is sought. Notice that the tangent of q is different from tan 22° 36′ by only 0.0004, and the total spread in the tangent values is 0.0020.

Therefore, the tangent of θ is $\frac{0.0004}{0.0020}$ of the way between the tangents of the two angles given in the table. This is 1/5 of the total spread, since

$$\frac{0.0004}{0.0020}=\frac{4}{20}=\frac{1}{5}$$

Another way of arriving at this result is to observe that the total spread is 20 ten-thousandths, and that the partial spread corresponding to angle q is 4 ten-thousandths. Since 4 out of 20 is the same as 1 out of 5, angle q is l/5 of the way between 22° 36′ and 22° 42′.

Taking 1/5 of the 6′ spread between the angles, we have the following calculation:

$$\frac{1}{5}\times 6'=\frac{1}{5}\times 5'60'' = 1'12'' \text{ (1 minute and 12 seconds)}$$

The 12" obtained in this calculation causes our answer to appear to have greater accuracy than the tables from which it is derived.

This apparent increase in accuracy is a normal result of interpolation. Final answers based on interpolated data should be rounded off to the same degree of accuracy as that of the original data.

The value of 1 minute and 12 seconds found in the foregoing problem is added to 22° 36′, as follows:

$$q = 22° \ 36' + 1'12'' = 22°37'12''$$

Therefore q is 22° 37′, approximately.

The foregoing problem could have been solved in terms of tenths and hundredths of a degree, rather than minutes, as follows:

$$\begin{array}{llll} & \textit{Angle} & \textit{Tangent} & \\ 0.1^\circ & \begin{cases} 22.60^\circ & 0.4163 \\ \theta & 0.4167 \\ 22.70^\circ & 0.4183 \end{cases} & 0.0020 & 0.0020 \end{array}$$

In this example, we are concerned with an angular spread of 0.10° and q is located 1/5 of the way through this spread. Thus we have

$$\theta = 22.60^\circ + \left(\frac{1}{5} \times 0.10^\circ\right)$$

$$\theta = 22.60^\circ + 0.20^\circ$$

$$\theta = 22.62^\circ$$

Interpolation must be approached with common sense, in order to avoid applying corrections in the wrong direction.

For example, the cosine of an angle decreases in value as the angle increases from 0° to 90°. If we need the value of the cosine of an angle such as 22°39′, the calculation is as follows:

$$\begin{array}{llll} & \textit{Angle} & & \textit{Tangent} \\ 6' & \begin{cases} 22^\circ 36' \\ 22^\circ 39' \\ 22^\circ 42' \end{cases} & 3' & \begin{matrix} 0.9232 \\ \\ 0.9225 \end{matrix} \; 0.0007 \end{array}$$

In this example, it is easy to see that 22°39′ is halfway between 22°36′ and 22°42′. Therefore the cosine of 22°39′ is halfway between the cosine of 22°36′ and that of 22°42′.

Taking one-half of the spread between these cosines, we then SUBTRACT from 0.9232 to find the cosine of 22°39′, as follows:

$$\cos 22^\circ 39' = 0.9232 - \left(\frac{1}{2} \times 0.0007\right) = 0.9232 - 0.00035$$

$$= 0.92285 = 0.9229 \text{ (approximately)}$$

Practice problems:

1. Use the table in appendix II to find the decimal value of each of the following ratios:

a. tan 45° d. sin 37°14'

b. sin 60° e. cos 51.5°

c. cos 42°6' f. tan 13.75°

2. Find the angle which corresponds to each of the following decimal values in appendix II:

a. sin θ = 0.2790 c. tan θ = 0.7604

b. cos θ = 0.9.18 d. sin θ = 0.8142

Answers:

1. a. 1 d. 0.6051

b. 0.8660 e. 0.6225

c. 0.7420 f. 0.2447

2. a. θ = 16.2° c. θ = 37°15'

b. θ = 25°36' d. θ = 54°30'

RIGHT TRIANGLES WITH SPECIAL ANGLES AND SIDE RATIOS

Three types of right triangles are especially significant because of 'their frequent occurrence. These are the 30°-60°-90° triangle, the 45°-90° triangle, and the 3-4-5 triangle.

THE 30°-60°-90° TRIANGLE

The 30°-60°-90° triangle is so named because these are the sizes of its three angles. The sides of this triangle are in the ratio of 1 to $\sqrt{3}$ to 2, as shown in figure 19- 10.

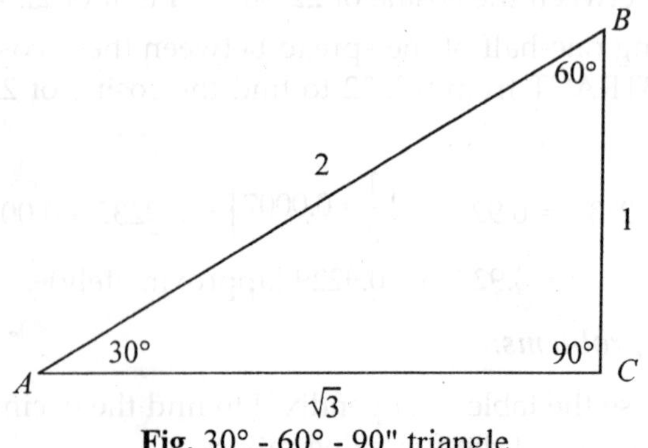

Fig. 30° - 60° - 90" triangle

The sine ratio for the 30° angle in figure establishes the proportionate values of the sides. For example, we know that the sine of 30° is 1/2; therefore side *AB* must be twice as long as *BC*. If side *BC* is 1 unit long, then side *AB* is 2 units long and, by the rule of Pythagoras, *AC* is found as follows:

$$AC = \sqrt{(AB)^2 - (BC)^2} = \sqrt{4-1} = \sqrt{3}$$

Regardless of the size of the unit, a 30°- 60°-90° triangle has a hypotenuse which is 2 times as long as the shortest side. The shortest side is opposite the 30° angle.

The side opposite the 60° angle is fl times as long as the shortest side. For example, suppose that the hypotenuse of a 30°-60°-90° triangle is 30 units long; then the shortest side is 15 units long, and the length of the side opposite the 60° angle is $15\sqrt{3}$ units.

Practice problems: Without reference to tables or to the rule of Pythagoras, find the following lengths and angles in figure:

1. Length of *AC*.
2. Size of angle *A*.
3. Size of angle *B*.
4. Length of *RT*.
5. Length of *RS*.
6. Size of angle *T*.

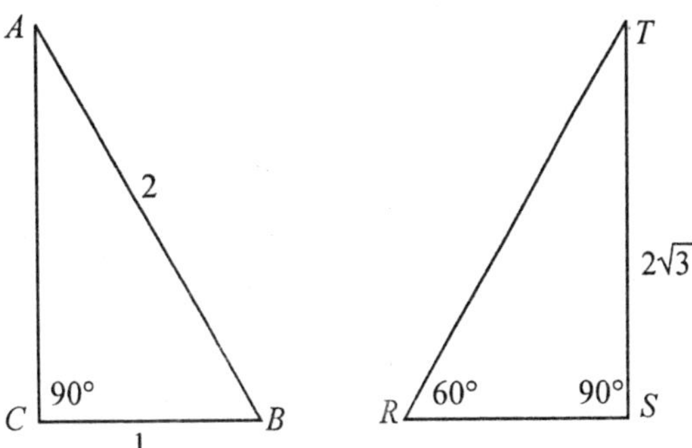

Fig. Finding parts of 30°-60°-90° Triangles

Answers:

1. $\sqrt{3}$
2. 30°
3. 60°
4. 4
5. 2
6. 30°

THE 45°-90° TRIANGLE

Figure illustrates a triangle in which two angles measure 45° and the third angle measures 90°. Since angles *A* and *B* are equal, the sides opposite them are also equal. Therefore, *AC* equals *CB*. Suppose that *CB* is 1 unit long; then *AC* is also 1 unit long, and the length of AB is calculated as follows:

$$(AB)^2 = 1^2 + 1^2$$

$$AB = \sqrt{2}$$

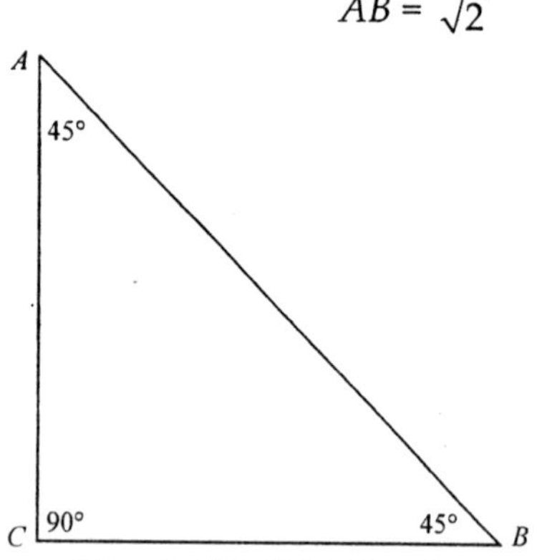

Fig. A 45°-90° Triangle

Regardless of the size of the triangle, if it has two 45° angles and one 90° angle, its sides are in the ratio 1 to 1 to $\sqrt{2}$. For example, if sides *AC* and *CB* are 3 units long, *AB* is $3\sqrt{2}$ units long.

Practice problems: Without reference to tables or to the rule of Pythagoras, find the following lengths and angles. in figure:

1. AB
2. BC
3. Angle B

Answers:

1. $2\sqrt{2}$
2. 2
3. 45°

THE 3-4-5 TRIANGLE

The triangle shown in figure has its sides in the ratio 3 to 4 to 5. Any triangle with its sides in this ratio is a right triangle.

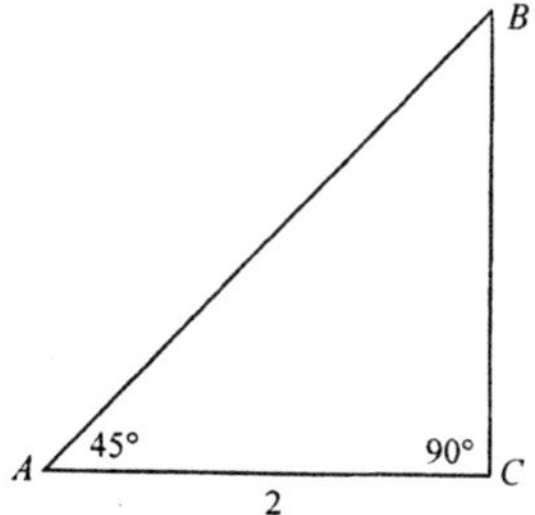

Fig. Finding unknown parts in a 45°-90° Triangle

It is a common error to assume that a triangle is a 3-4-5 type because two sides are known to be in the ratio 3 to 4, or perhaps 4 to 5. Figure 19-15 shows two examples of triangles which happen to have two of their sides in the stated ratio, but not the third side. This

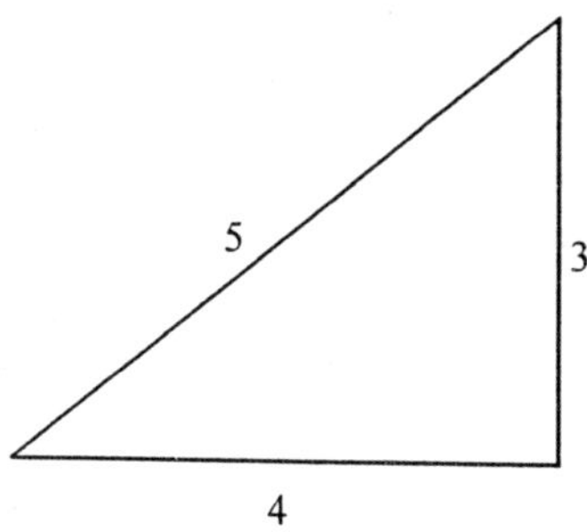

Fig. A 3-4-5 Triangle

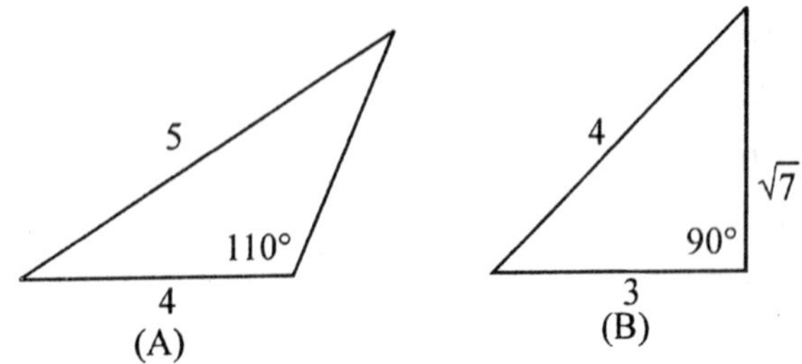

Fig. Triangles which may be Mistaken for 3-4-5 Triangles

can be because the triangle is not a right triangle, as in figure. On the other hand, even though the triangle is a right triangle its longest side may be the 4-unit side, in which case the third side cannot be 5 units long.

It is interesting to note that the third side in figure 19-15 (B) is $\sqrt{7}$. This is a very unusual coincidence, in which one side of a right triangle is the square root of the sum of the other two sides.

Related to the basic 3-4-5 triangle are all triangles whose sides are in the ratio 3 to 4 to 5 but are longer (proportionately) than these basic lengths. For example, the triangle pictured in figure 19-6 is a 3-4 -5 triangle.

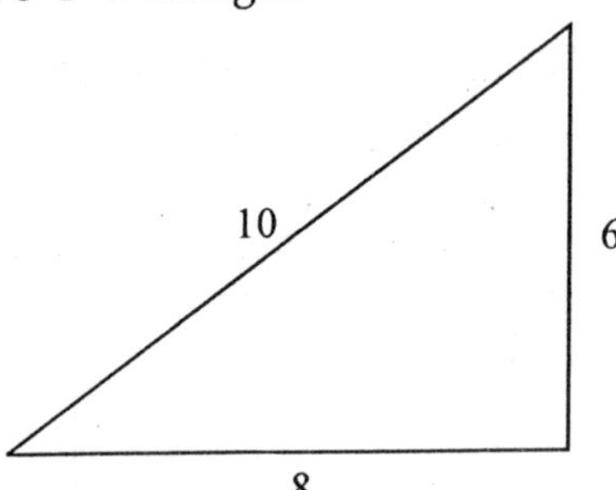

Fig. Triangle with sides which are Multiples of 3, 4, and 5

The 3-4-5 triangle is very useful in calculations of distance. If the data can be adapted to fit a 3-4-5 configuration, no tables or calculation of square root (Pythagorean Theorem) are needed.

Example: An observer at the top of a 40-foot vertical tower knows that the base of the tower is 30 feet from a target on the ground. How does he calculate his slant range (direct line of sight) from the target?

Solution: Figure shows that the desired length, *AB*, is the hypotenuse of a right triangle whose shorter sides are 30 feet and 40 feet long. Since these sides are in the ratio 3 to 4 and angle *C* is 90°) the triangle is a 3-4-5 triangle. Therefore, side *AB* represents the 5-unit side of the triangle. The ratio 30 to 40 to 50 is equivalent to 3-4-5, and thus side *AB* is 50 units long.

Practice problems: Without reference to tables or to the rule of Pythagoras, solve the following problems:

1. An observer is at the top of a 30-foot vertical tower. Calculate his slant range from a target on the ground which is 40 feet from the base of the tower.

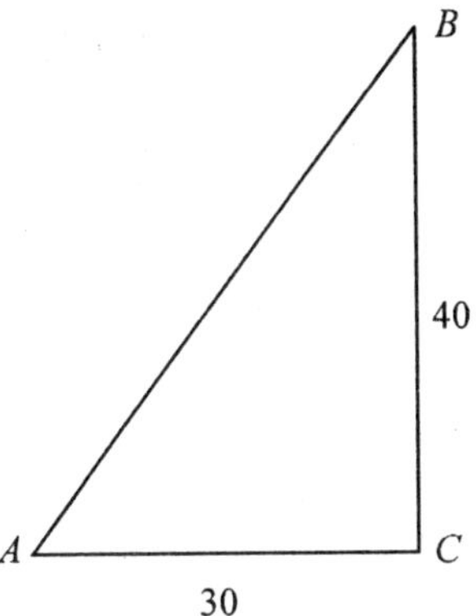

Fig. Solving Problems with a 3-4-5 Triangle

2. A guy wire 15 feet long is stretched from the top of a pole to a point on the ground 9 feet from the base of the pole. Calculate the height of the pole.

Answers:

1. 50 feet
2. 12 feet

OBLIQUE TRIANGLES

Oblique triangles were defined in chapter of this training course as triangles which contain no right angles. A natural approach to the solution of problems involving oblique triangles is to construct perpendicular lines and form right triangles which subdivide the original triangle. Then the problem is solved by the usual methods for right triangles.

DIVISION INTO RIGHT TRIANGLES

The oblique triangle ABC in figure has been divided into two right triangles by drawing line BD perpendicular to AC. The length of AC is found as follows:

1. Find the length of AD.

$$\frac{AD}{35} = \cos 40°$$

$$AD = 35 \cos 40°$$

$$= 35\ (0.7660)$$

$$= 26.8 \text{ (approximately)}$$

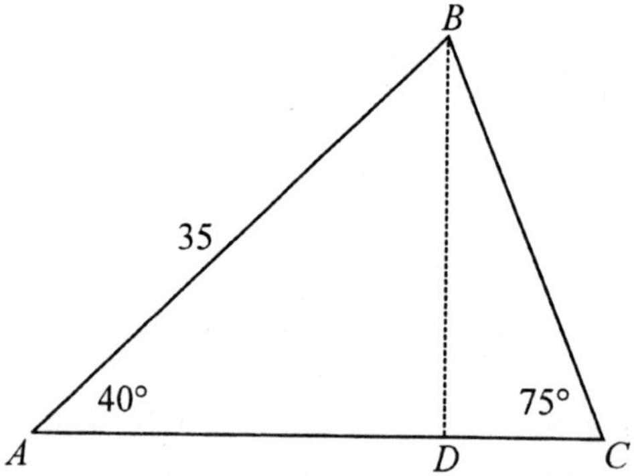

Fig. Finding the unknown parts of an Oblique Triangle

Caution: A careless appraisal of this problem may lead the unwary trainee to represent the ratio AC/AB as the cosine of 40°. This error is avoided only by the realization that the trigonometric ratios are based on Right triangles.

2. In order to find the length of *DC*, first calculate *BD*.

$$\frac{BD}{35} = \sin 40°$$

$$BD = 35 \sin 40°$$

$$= 35\ (0.6428)$$
$$= 22.4 \text{ (approximately)}$$

3. Find the length of DC

$$\frac{22.4}{DC} = \tan 75°$$

$$DC = \frac{22.4}{\tan 75°} = \frac{22.4}{3.732}$$

$$= 6.01 \text{ (approximately)}$$

4. Add *AD* and *DC* to find *AC*.

$$26.8 + 6.01 = 32.81$$

$$AC = 32.8 \text{ (approximately)}$$

SOLUTION BY SIMULTANEOUS EQUATIONS

A typical problem in trigonometry is the determination of the height of a point such as B in figure.

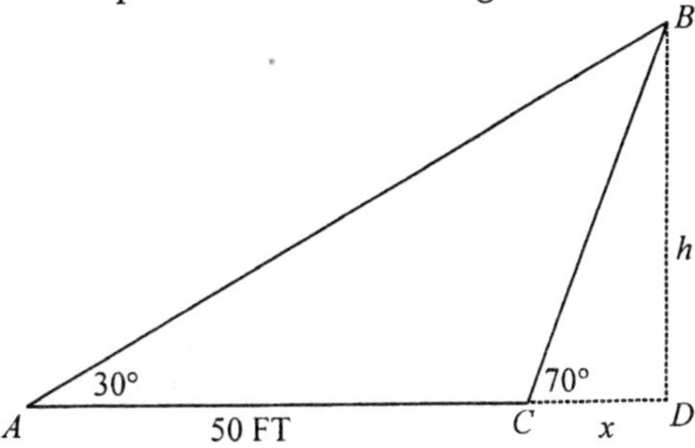

Fig. Calculation of unknown Quantities by means of Oblique Triangles

Suppose that point *B* is the top of a hill, and point *D* is inaccessible. Then the only measurements possible on the ground are those shown in figure. If we let *h* represent *BD* and *x* represent *CD*, we can set up the following system of simultaneous equations:

$$\frac{h}{x} = \tan 70°$$

$$\frac{h}{50+x} = \tan 30°$$

Solving these two equations for h in terms of x, we have

$$h = x \tan 70°$$

and

$$h = (50 + x) \tan 30°$$

Since the two quantities which are both equal to h must be equal to each other, we have

$$x \tan 70° = (50 + x) \tan 30°$$

$$x(2.748) = (0.5774) + x(0.5774)$$

$$x\,(2.748) - x\,(0.5774) = 28.8$$

$$x\,(2.171) = 28.8$$

$$x = \frac{28.8}{2.171} = 13.3 \text{ feet}$$

Knowing the value of x, it is now possible to compute h as follows:

$$h = x \tan 70° = 13.3\,(2.748)$$

$$= 36.5 \text{ feet (approximately)}$$

LAW OF SINES

The law of sines provides a direct approach to the solution of oblique triangles, avoiding the necessity of subdividing into right triangles. Let the triangle in figure (A) represent any oblique triangle with all of its angles acute. The labels used in figure are standardized. The small letter a is used for the side opposite angle A; small b is opposite angle B; small c is opposite angle C.

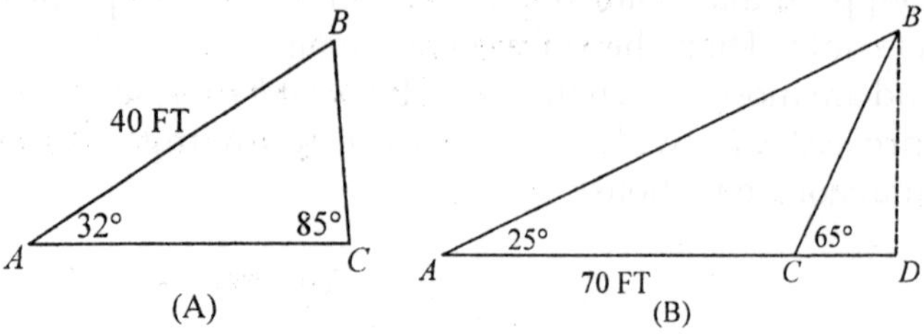

Fig. (*A*) Oblique Triangle with all Angles Acute; (*B*) Obtuse Triangle

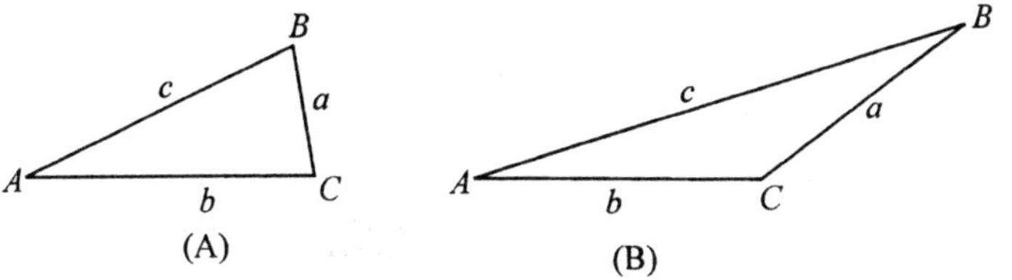

Fig. (*A*) Acute Oblique Triangle with Standard labels; (*B*) Obtuse Triangle with Standard Labels

The law of sines states that in any triangle, whether it is acute as in figure (*A*) or obtuse as in figure (*B*), the following is true:

$$\frac{a}{\sin A} = \frac{b}{\sin B} = \frac{c}{\sin C}$$

Example: In figure (*A*), let angle *A*-be 15° and let angle C be 85°. If *BC* is 20 units, find the length of *AB*.

Solution: By the law of sines,

$$\frac{20}{\sin 15°} = \frac{c}{\sin 85°}$$

$$c = \frac{20 \sin 85°}{\sin 15°}$$

$$c = \frac{20\,(0.9962)}{0.2588} = 77.0$$

Index

A

Absolute value 32, 33, 78, 112, 148, 277
Accuracy 21, 52, 54, 55, 56, 58, 59, 65, 113, 128, 217, 286, 292
Adding 3, 5, 6, 10, 12, 21, 33, 38, 42, 58, 59, 61, 65, 88, 98, 101, 184, 185, 195, 196, 218, 256
Addition method 218, 219, 221, 223
Algebraic expressions 137, 140, 141, 146, 179
Algebraic Sums 139
Associative Law 43, 44, 139
Axioms 42, 43, 192, 194

B

Binomial 141, 142, 153, 155, 157, 158, 163, 168, 169, 170, 171, 172, 173, 174, 177, 246, 255
Binomial Forms 172

C

Carry and Borrow 7
Combined Multiplication 85
Combining Radicals 87
Common Errors 74
Common Factors 166, 168, 179
Common Logarithms 102, 103, 104
Common term 156, 159, 160, 177
Commutative law 44, 138, 149, 169
Complex numbers 238, 239, 242, 243, 244, 245, 246, 247, 268
Complex Plane 238, 239
Components of Logarithms 107
Conditional Equations 191
Conjugates 245, 246
Constants and Variables 188
Coordinate Axes 210
Coordinate System 31, 209, 210, 228, 237, 263
Cube root 67, 68, 80, 96, 99, 112, 133, 134, 135, 136

D

Decimal points 52, 121, 122, 123
Decimals to Percent 47
Degree of an Equation 189, 249
Denominate Numbers 21, 22, 88, 143, 151, 152
Denominators 46, 94, 148, 157, 181, 184, 247
Discriminant 267, 268, 269, 270, 271
Distributive law 43, 45, 139, 153
Dividing Fractions 182
Dividing Monomials 149
Division Combined 123
Division Methods 17
Double Roots 268, 269

E

End Zeros 15, 51, 52, 127, 129

Equality 1, 42, 187, 188, 189, 190, 191, 192, 200, 204, 253, 269
Estimation 19, 52
Evaluating Radicals 96
Even Exponents 172

F

Factoring Radicals 90
Fractional Exponents 81, 86, 93, 106
Fractions 5, 12, 16, 22, 41, 42, 46, 47, 54, 94, 179, 181, 182, 184, 185, 187, 200, 230, 283
Function value 283

G

Graphical Interpretation 272, 273
Graphical Solution 216, 264
Grouping Symbols 145, 146, 187

I

Identities 190
Imaginary Numbers 93, 230, 232, 233, 235, 237, 244
Imaginary Roots 268, 273
Imaginary Unit 233, 234, 238, 243
Inequalities 204, 205, 208, 227, 229
Interpolation 285, 286
Irrational Numbers 94, 99, 231, 260, 272
Irrational Roots 270

L

Law of Sines 296, 297
Laws of Exponents 73, 74, 75, 77, 82
Like Signs 33, 34, 37, 38, 39
Linear Equations 188, 208, 262
Lines Parallel 224
Literal Coefficients 197, 222
Literal Exponents 168

M

Mental Multiplication 158
Micrometer 60, 62, 65, 66
Micrometers and Verniers 60
Missing term 175, 176
Monomial 141, 142, 148, 153, 154, 155, 168, 169
Multiplication and Division 85, 90, 118, 206, 242
Multiplying Fractions 181
Multiplying Monomials 148

N

Natural Logarithms 102, 104
Negative Exponents 78, 148, 150
Negative Fractional 106
Negative Integers 69
Negative Logarithms 106
Number line 4, 5, 33, 34, 35, 36, 208, 232, 235
Number Set 1

O

Oblique triangles 294, 296
Odd Exponents 172
One Operation 194

P

Parentheses 11, 12, 40, 41, 137, 141, 144, 145, 146, 147, 152, 154, 161, 167, 175, 177, 199, 210, 222, 242
Partial products 14, 15
percent of error 56, 58
Percent to a Decimal 48, 49
Points and Lines 31
Polynomials 141, 151, 152, 155, 163, 166, 179, 181, 187, 242

Positioning Numbers 127
Positive Fractional 105
Positive Integers 1, 2, 76, 77, 251
Powers 67, 73, 75, 82, 83, 84, 85, 86, 91, 92, 93, 96, 101, 103, 106, 107, 108, 161, 163, 165, 234, 237, 244, 249, 254
Prime Factors 25, 26, 27, 93, 167, 184
Pythagorean Theorem 277, 278, 284, 292

Q

Quadrants 211
Quadratic Equations 249, 251, 273, 274

R

Radicals 86, 88, 89, 90, 91, 93, 94, 99, 147, 148, 260
Rational 94, 95, 96, 177, 231, 269, 270, 271, 272
Rational Roots 270
Rationalizing 157, 247
Reducing Fractions 179
Regrouping 9, 10, 17, 18
Removing Parentheses 145, 146, 152
Right Triangles 276, 278, 279, 288, 294, 296

S

Scientific Notation 82, 83, 85, 108
Sense Reversal 207
Side Ratios 288
Signed Numbers 29, 31, 33, 35, 37, 39
Significant Digits 56, 57, 58, 59, 85, 113, 117, 119, 121, 128
Slide Rule Theory 113
Special Exponents 76
Special Productst 156
Square Roots 98, 134, 170, 174, 175, 232, 235, 252, 255, 256
Subsets 4, 5
Substitution Method 221, 222
Subtracting 8, 42, 58, 88, 143, 151, 184, 195, 199
Subtraction and Division 195
Sum and Difference 156, 157, 158, 170, 172, 246
Symbols of Grouping 144, 145, 146, 147, 152
Synthetic Division 164, 165

T

Terms and Coefficients 192
Tests For Divisibility 27
Thermometer 4, 30, 31
Trigonometric Ratios 281, 283, 294
Trinomial Squares 175
Two Conjugates 246
Two Squares 170, 171, 172
Two Variables 215, 225, 226, 227
Two Variables 190, 208, 209, 211, 213, 215, 223, 226, 227

U

Unequal Roots 270
Uneven Division 20
Unlike Signs 33, 34, 37, 39

V

Venire 63
Verbal Problems 273

W

Word Problems 225